Recent Progress in Arithmetic and Algebraic Geometry

CONTEMPORARY MATHEMATICS

386

Recent Progress in Arithmetic and Algebraic Geometry

Barrett Lecture Series Conference
April 25–27, 2002
University of Tennessee
Knoxville, Tennessee

Yasuyuki Kachi
S. B. Mulay
Pavlos Tzermias
Editors

American Mathematical Society
Providence, Rhode Island

This volume consists of the thirty-first annual John H. Barrett Memorial Lecture Series, held April 25–27, 2002 at the University of Tennessee, Knoxville, with financial support from the University of Tennessee, the Science Alliance, and the Haines-Morris Foundation.

2000 *Mathematics Subject Classification.* Primary 14–XX.

Library of Congress Cataloging-in-Publication Data

Barrett Lecture Series Conference (31st : 2002 : University of Tennessee, Knoxville)

Recent progress in arithmetic and algebraic geometry : Barrett Lecture Series Conference, April 25–27, 2002, University of Tennessee, Knoxville, Tennessee / Yasuyuki Kachi, S. B. Mulay, Pavlos Tzermias, editors.

p. cm. — (Contemporary mathematics, ISSN 0271-4132 ; 386)

Includes bibliographical references.

ISBN 0-8218-3401-0 (alk. paper)

1. Geometry, Algebraic—Congresses. 2. Arithmetic—Congresses. I. Kachi, Yasuyuki, 1969– II. Mulay, S. B., 1956– III. Tzermias, Pavlos, 1968– IV. Title. V. Contemporary mathematics (American Mathematical Society) ; v. 386.

QA564.B36 2002

516.3′5—dc22 2005052406

∞ The paper used in this book is acid-free and falls within the guidelines
established to ensure permanence and durability.
Visit the AMS home page at http://www.ams.org/

10 9 8 7 6 5 4 3 2 1 10 09 08 07 06 05

Contents

Preface

This volume consists of the proceedings of the thirty-first annual John H. Barrett Memorial Lecture Series, held April 25-27, 2002 at The University of Tennessee, Knoxville. The lecture series was on the twin research areas : Algebraic Geometry / Arithmetic Algebraic Geometry. There were three one-hour plenary lectures by two Principal Lecturers. For the Algebraic Geometry part, Yujiro Kawamata of University of Tokyo was the Principal Lecturer. For the Arithmetic Algebraic Geometry part, Yasutaka Ihara of the Kyoto University Research Institute for Mathematical Sciences was the Principal Lecturer. In addition to the six plenary lectures, there were twelve forty-minute talks delivered by the participating guest lecturers coming from various other universities.

We would like to thank the Principal Lecturers for gracing the conference with their inspiring plenary lectures and also like to thank all the participants for making the lecture series a stimulating event. We would like to thank all the contributors to this volume. Our sincere thanks are due to The University of Tennessee, The Science Alliance and the Haines-Morris Foundation for their financial support of the lecture series. We would like to thank John B. Conway, the program director of NSF, and Robert J. Daverman, the Secretary of AMS, for their consistent support.

We would like to thank the AMS Acquisitions, the Managing Editor Dennis DeTurck for the approval, the Associate Publisher Sergei Gelfand for helping the timely publication, and the Editorial Assistant Christine M. Thivierge for the exclusive encouragement, assistance and cooperation, with much patience.

Yasuyuki Kachi

S. B. Mulay

Pavlos Tzermias

John H. Barrett (1922–1969)

John H. Barrett was born in 1922 and grew up on a farm in Kansas. He received his A. B. in 1944 from Fort Hays Kansas State College. He did his graduate work at the University of Texas and was an instructor there from 1946–1951. His thesis was on differential equations of non-integer order and under the direction of H. J. Ettlinger. He was an assistant professor at Delaware from 1951–1956, an associate professor at Utah from 1956–1961, and professor at Tennessee from 1961–1969. He also spent the year 1955–1956 visiting Yale and the year 1959–1960 visiting the Research Center at Wisconsin. In January, 1969, he died from complications following a kidney transplant. His wife, Lida, was also a faculty member at Tennessee, where she was head from 1973 to 1980. They had three children.

John was well known for his work on oscillation and disconjugacy theory of linear differential equations as well as the study of boundary value problems. One of his best known contributions was the extension of the classical Prüfer transformation for 2nd order scalar differential equations to systems of differential equations. He had several Ph. D. students. In a time when a common teaching load was three courses, he regularly ran a seminar on differential equations in addition. Typically it would meet very early on a Tuesday or Thursday morning with John as the most frequent lecturer. Through his efforts and influence, an active group in differential equations evolved at the University of Tennessee. This tradition has continued since that time. The Barrett Lecture series is a fitting reminder of his many contributions to the university and community.

(A tribute delivered by Don Hinton, March 22, 1995)

List of Invited Speakers

Yasutaka Ihara (Plenary Speaker)
Kyoto University

Yujiro Kawamata (Plenary Speaker)
University of Tokyo

Steven Dale Cutkosky
University of Missouri

Jordan Ellenberg
Princeton University

Farshid Hajir
California State University at San Marcos

Minhyong Kim
University of Arizona

Sándor J. Kovács
University of Washington

Tie Luo
University of Texas at Arlington

Kenji Matsuki
Purdue University

William McCallum
University of Arizona

Keiji Oguiso
University of Tokyo

Rachel J. Pries
Columbia University

Romyar Sharifi
Harvard University

Qi Zhang
University of Missouri

Contemporary Mathematics
Volume **386**, 2005

COMPACT KÄHLER THREEFOLDS OF π_1-GENERAL TYPE.

F. CAMPANA AND Q. ZHANG

§0. Introduction

Let X be a compact complex manifold, connected and of complex dimension n. Let $\kappa(X)$ be its Kodaira dimension, and $\alpha_X : X \to Alb(X)$ its Albanese map. A fibration $f : X \to Y$ is a surjective meromorphic map with connected fibres.

Let $\widetilde{X}$ be the universal cover of X. We say that X is of π_1*-general type* if $\widetilde{X}$ is not the union of its positive-dimensional compact analytic subsets.

Obvious examples of such $X's$ are complex tori and compact quotients of bounded domains in $\mathbb{C}^n$. We conjecture that the general case is essentially a simple combination of these:

Conjecture 0: *Let X be a compact Kähler manifold of π_1-general type. Then:*
1. $\kappa(X) \geq 0$.
2. There exists a finite étale cover of X which is bimeromorphic to a manifold X' such that its Iitaka-Moishezon fibration[1] $f : X' \to Y'$ *is a submersion with fibres complex tori, and with base Y' a projective manifold of general type and of π_1-general type. (f being a submersion means that it is holomorphic and of maximal rank $dim(Y')$ everywhcrc on X')*

This conjecture extends the one made in [C95]: *if X is of π_1-general type, and if $\kappa(X) = 0$, then X is bimeromorphically covered by a torus.* Due to [K-O], the above conjecture also extends the one by S.Iitaka ([I]: *if $\widetilde{X} \cong \mathbb{C}^n$, then X is covered by a torus).*

See §4. below for a motivation of this conjecture, and of the terminology.

Notice that the above conjecture already appears in [K] for X projective.

For curves , the conjecture holds by the uniformization theorem. The conjecture can easily be shown for surfaces as well, using the Enriques-Kodaira-Shafarevich classification.

We show here that the conjecture holds for threefolds also.

More precisely:

Theorem 0: *The above conjecture holds in any of the following three situations:*
1. X is projective, K_X is nef, and the Abundance conjecture holds (ie: the pluricanonical system $|mK_X|$ is free for some $m > 0$).

[1] usually called the Iitaka fibration. See however [Moi].

2. X is a projective threefold.

3. X is a compact non-projective Kähler threefold.

In the projective case, we just have to tie up together results already well-known, among which [K,6.3] (which solves the case where the generic fibre of f is an Abelian variety) plays a crucial role. For non-projective threefolds, we need to use the results of [C95] to rule out the (conjecturally non-existing) simple non-Kummer case. We indeed show in arbitrary dimension that $\chi(\mathcal{O}_X) \neq 0$, if X is not of general type, but of π_1-general type, which implies in dimension three that either $p_g \geq 2$, or $q \geq 1$.

The results of [CP] immediately rule out the case $\kappa(X) = -\infty$, because they then imply the uniruledness of X, and solve the case $\kappa(X) = 0$, because they then imply that if X is not simple, it is bimeromorphically and after a finite étale cover, the product of an elliptic curve with a $K3$ surface.

To deal with the remaining cases $\kappa(X) = 1, 2$, we study the Iitaka-Moishezon fibration. One can reduce to the case where its generic fibre is a complex torus, and adapt most of the arguments of [K,6.3]. We actually obtain a more precise result, in this case:

Theorem 1: *Let X be a compact Kähler threefold of π_1-general type. Then:*

1. $\kappa(X) \geq 0$, and a suitable finite étale cover X' of X has $f' : X' \to Y'$, its Iitaka-Moishezon fibration, bimeromorphic:

2. Either: to $Alb(X') \times_{Alb(Y')} Y'$, and Y' is of general type and of π_1-general type.

3. Or: to a submersion with all fibers Abelian surfaces, and base Y' a curve of genus 2 or more.

Notice that the proof uses properties very specific to the three-dimensional case.

From this and the arguments of the proof (especially the proof of 5.5), we immediately deduce a concrete description of such threefolds (see §6).

Corollary 2: *Let X be a compact Kähler threefold of π_1-general type. Then, after a finite étale cover, X is bimeromorphic to one of the following:*

1. A complex torus.

2. Either:

2.a. A subvariety of Kodaira dimension 1 of a complex torus, or:

2.b. A threefold with a submersion onto a curve of general type with Abelian surfaces as fibers.

3. A variety V of type 3.a, 3.b, or 3.c below:

3.a. V has a finite map onto a threefold of Kodaira dimension 2 of a complex torus.

3.b. $V = E \times S$, where E is an elliptic curve, and S a surface of general type and π_1-general type such that $q(S') = 0$, for any finite étale cover S' of S.

3.c. $V = S \times_C S'$, where S, S' are two surfaces fibered over a curve C of general type, with S of general type and π_1-general type, and S' a surface in a complex torus, elliptic over C.

4. A variety of general type.

In particular, S.Iitaka's conjecture is true for compact Kähler threefolds (this is shown also in [N"], some intermediate results of which are similar to some of ours).

We now prove Theorems 0 and 1, according to the three cases of Theorem 0.

§1. X PROJECTIVE, K_X NEF, ASSUMING ABUNDANCE.

Because X is of π_1-general type (or because K_X is nef), it is not uniruled. In particular, if the abundance conjecture holds for X, then the Kodaira dimension $\kappa(X)$ of X should be nonnegative.

First, let us assume that $\kappa(X) = 0$, because K_X is nef, from Abundance we deduce that K_X is torsion. Calabi-Yau and a result of Beauville and Bogomolov [B] tell us that (after a finite étale cover), X will split as the product of some manifold with a simply-connected Calabi-Yau manifold. Because X is of π_1-general type, the second factor is a point, and X is covered by a complex torus.

If X is of general type, there is nothing to prove.

Now, let us assume that: $n > \kappa(X) > 0$. Then Abundance conjecture assures that there exists a morphism (the Iitaka-Moishezon fibration) $f : X \to Y$. Its generic fibre is smooth with torsion canonical bundle. By the previous step, this generic fiber is an Abelian variety. From [K, 6.3], we get the conclusion.

§ 2. PROJECTIVE THREEFOLDS.

The above arguments work for this case, because the Abundance conjecture is known in this case, by the work of Kawamata, Miyaoka, Mori and Shokurov.

We just need to deal with the case $\kappa(X) = 0$, X of π_1-general type. Then [N-S] shows that a finite étale cover of X is bimeromorphic to a torus.

§ 3. KÄHLER THREEFOLDS.

Let now X be a compact Kähler (connected smooth) threefold of π_1-general type with $\kappa(X) \in \{-\infty, 0, 1, 2\}$.

We now make a case-by-case study, according to the value of $\kappa(X)$.

We settle in this section the cases where $\kappa(X)$ is either 0 or $-\infty$.

3.1: Case $\kappa(X) = -\infty$.

Then X is not projective, by §2. This case is also excluded in the non-projective case by [CP,8.1]. Indeed, by [C-P], we get that X is either uniruled or "simple" (which means: not covered by compact analytic subsets of intermediate dimensions either 1 or 2). In particular, if X is simple, the algebraic dimension $a(X)$ equals zero (ie: all global meromorphic functions on X are constant), and there is no surjective meromorphic fibration $f : X \to Y$ with Y a curve or a surface. This "simple" case is excluded by the following lemma, since we assumed $\kappa(X) = -\infty$:

3.1.1 Lemma: *Let X be a simple compact Kähler threefold. If X is of π_1-general type, then X is bimeromorphic to a torus. (Recall that a threefold X is simple if it is not the union of its compact curves and surfaces).*

Proof: From theorem 3.1.2 below, we learn that $\chi(\mathcal{O}_X) = 0$. Because $h^{2,0} \geq 1$, by the result of Kodaira recalled above, we see that $q + p_g \geq 2$, and so either $p_g := h^{0,3}(X) \geq 2$, or $q := h^{1,0} \geq 1$. The first case is excluded because it would produce a nonconstant meromorphic function on X. In the second case, the Albanese map $\alpha_X : X \to Alb(X)$ needs to be surjective and to have connected fibers (because $a(X) = 0$; see [U]). Because X is moreover simple, its fibres are zero-dimensional (because $q = dim(Alb(X)) > 0$). Thus α_X is bimeromorphic.

We have used the following result, whose proof is given in 4.3.4):

3.1.2 Theorem: *Let X be a compact Kähler n-fold of π_1-general type. Then: either X is projective and of weak general type (ie: $\kappa^+(X) = n$, see §4.), or: $\chi(\mathcal{O}_X) = 0$.*

3.2 Case $\kappa(X) = 0$.

In the projective case, the conclusion follows from §2. This case is easily solved again using [CP, 8.1] in the non-projective case: X has then to be either simple, or after a finite étale cover, bimeromorphic to either a torus, or the product of an elliptic curve with a $K3$ surface. Since X is of π_1-general type, X is bimeromorphic to a torus.

3.3 Reduction steps when $\kappa(X) \geq 1$.

We then assume that $\kappa(X) = 1$ or 2. Let $f : X \to Y$ be the (holomorphic, after resolution of its indeterminacies) Iitaka-Moishezon fibration of X. Its base Y is either a curve or a surface. Its generic fibre X_y has $\kappa(X_y) = 0$, and is of π_1-general type. Thus X_y is an elliptic curve if $\kappa(X) = 2$, and bimeromorphic to either a torus or a bielliptic surface if $\kappa(X) = 1$. In this last case, by a finite étale cover of X, we can assume X_y to be bimeromorphic to a torus, by using a very special case of [K, 6.4.1] . We thank N.Nakayama who pointed out a gap in the proof given in the previous version of the present text, where a more general version was stated. The special case below is the same one as in [N", 6.4].

3.3.0. Proposition: *Let $H < G$ be groups, with H normal in G, such that H is finitely generated and residually finite. Let H_1 be an abelian characteristic subgroup of finite index in H. Then:*

1. G has a normal subgroup G' of finite index such that $H' := G' \cap H \subset Z.H_1$, where $Z \subset H$ is the centraliser of H_1 in H (ie: the subgroup of H consisting of all elements commuting with each element of H_1). Moreover, G' can also be chosen such that the quotient group $(G' \cap Z.H_1)/(G' \cap H_1)$ is abelian.

2. If H_1 is abelian and torsionfree, then G' can be chosen such that $(G' \cap H)$ is abelian.

Proof: Of (1): Let $F := H/H_1$, and $c : H \to Out(H_1) := Aut(H_1)/Int(H_1)$ be the action by conjugation. Its kernel is $Z.H_1$, and its image F' is a quotient of F, hence finite, because c is trivial on H_1.

For K a characteristic subgroup of finite index of H_1, define, by using conjugation again: $c_K : H \to Aut(H_1/K)$. Because H is residually finite, and H_1 is abelian, we can choose K such that the kernel of c_K is included in the kernel of c.

Indeed, if $h \in H$ does not belong to the kernel of c, then $[h, a] := hah^{-1}a^{-1} \neq 1$, for some $a \in H_1$. Thus $[h, a] \notin K$, for some suitable K, characteristic and of finite index in H_1, by the residual finiteness of H. And so, h does not act trivially on H_1/K', for any smaller $K' \subset K$, as desired. Because F' is finite, we need to check this condition for finitely many $h's$ in H, say for $h_i, i \in H/Z.H_1$, representatives of $H/Z.H_1$, since if $h = h_i.z.b$, for $z \in Z, b \in H_1$, we have: $[h, a] = [h_i, a].[b, a] = [h_i, a] \neq 1$, because $z \in Z$, and H_1 is abelian. Hence the claim.

We choose and fix K such that the preceding condition is satisfied.

Define now $r : G \to Aut(H_1/K))$ by conjugation. Because the image group is finite, the kernel $G"$ of r is of finite index in G. And $G" \cap H \subset Z.H_1$, by the above construction and observation. To ensure the second condition, we simply need to ask G' to be the kernel of the conjugation map: $r' : G" \to Aut(H/H_1)$.

Proof of (2): this follows from the following standard claim (applied to $A = (G' \cap H_1)$, and $B := G' \cap Z.H_1$):

Claim: Let $A \subset B$ be groups, with A central (hence abelian), torsionfree of finite index in B, and such that B/A is abelian. Then B is abelian.

Proof: For $x, y, z \in B$, we have :$xyx^{-1}y^{-1} := [x, y] = [xA, yA]$, and by easy commutator calculus: $[x, yz] = [x, y][x, z]^y = [x, y][x, z]$ (since B acts trivially on A). Thus, for any $x \in B$, $[x, B]$ is a finite subgroup of A. And is thus trivial, by the torsionfreeness of A.

This being done, we can use [C85] (the construction of relative Albanese maps) to further reduce to the special case when X_y is a torus.

We can and shall thus assume in the sequel that X_y is a torus for $y \in Y$ generic.

3.3.1 Lemma: *Assume that $\kappa(X) = 1$, for X a compact Kähler threefold of π_1-general type. Let $f : X \to Y$ be its Iitaka-Moishezon fibration. Then Y is a curve of general type (ie: $g(Y) \geq 2$), after X has been replaced by a suitable finite étale cover.*

Hence we can (and shall) assume that f has no multiple fibres, and that Y is a curve of general type, if $\kappa(X) = 1$.

Proof: Assume the contrary. After replacing X by a suitable finite étale cover, we can assume that f is π_1-exact, which means that the natural sequence of groups (3.3.1.1) below is exact:

(3.3.1.1) $\pi_1(X_y) \to \pi_1(X) \to \pi_1(Y) \to 1$.

This property is shown in [C98], and is peculiar to the case of 1-dimensional bases. Thus $\pi_1(X)$ is solvable of finite rank, and the Albanese map of X and any of its finite étale covers is obviously surjective (because no such cover has a surjective map onto a positive dimensional variety of general type). In this case, it follows from [C01] and [C95'] that $\pi_1(X)$ is almost abelian (ie: has a subgroup of finite index which is abelian).

By taking a suitable finite étale cover of X, we shall thus assume $\pi_1(X)$ to be abelian, and moreover that $\alpha_X : X \to Alb(X)$, the Albanese map of X induces an isomorphism of fundamental groups. Because X is of π_1-general type, α_X has to be generically finite. Because the fibres of f are tori, the restriction of α_X to X_y has to be étale onto its image, and so isomorphic, because α_X induces an isomorphism of fundamental groups. Thus α_X is bimeromorphic, and $\kappa(X) = 0$, in contradiction with our hypothesis.

3.3.2 Proposition: *Let X be a compact kähler threefold of π_1-general type, with $\kappa(X) \geq 1$. Let $f : X \to Y$ be its Iitaka-Moishezon fibration.*

Replacing X by a suitable finite étale cover, X is either:

bimeromorphic to $Alb(X) \times_{Alb(Y)} Y$, or such that, after taking a suitable bimeromorphic model, f is a submersion onto the curve of general type Y with Abelian surfaces as fibres.

Proposition 3.3.2, which is the main assertion of theorem 1, will be proved in §5.

The proof of theorem 1 is then completed by means of the easy:

3.3.3 Lemma: *Let $f : X \to Y$ be the Iitaka-Moishezon fibration of X kähler compact. Assume that X is bimeromorphic to $Alb(X) \times_{Alb(Y)} Y$. Then:*

1. $\kappa(X) = \kappa(Y)$.

2. X is of π_1-general type if and only if so is Y. More generally:

3. $dim(X) - \gamma d(X) = dim(Y) - \gamma d(Y)$ (see §4. for this notion).

Proof: Assertion 1 follows from the equality:

$K_{X/Y} = \alpha_X^*(K_{[Alb(X)/Alb(Y)]}) = \alpha_X^*(\mathcal{O}_{Alb(X)})$.

For assertion 3, we have a natural map h:

$h : \Gamma(X) \to Z := \Gamma(Alb(X)) \times_{\Gamma(Alb(Y))} \Gamma(Y) = Alb(X) \times_{Alb(Y)} \Gamma(Y)$, (induced from the equality: $X = Alb(X) \times_{Alb(Y)} Y$).

We shall show that h is generically finite. Let $H \subset X$ be an irreducible compact analytic subset contained in a general fibre of the composite map $h \circ \gamma_X : X \to Z$. We have to show that its image in $\Gamma(X)$ is point. But this is immediate, because H is of the form:$a \times G$, for $G \subset Y$ a general fibre of γ_Y. The image of $\pi_1(\hat{H})$ in $\pi_1(X)$ is thus finite. This shows assertion 3, of which assertion 2 is a special case.

§4. VANISHING OF $\chi(\mathcal{O}_X)$.

We recall here the two bimeromorphic invariants $\kappa^+(X)$ and $\gamma d(X)$, introduced in [C95] for compact n-dimensional Kähler manifolds X.

4.1 The invariant $\gamma d(X)$.

Let $r : \widetilde{X} \to X$ be the universal cover. There exists a unique proper connected fibration $\tilde{\gamma} : \widetilde{X} \to \widetilde{\Gamma}(X)$ such that if $\tilde{x} \in \widetilde{X}$ is a general point, any connected compact analytic subset of $\widetilde{X}$ going through $\tilde{x}$ is contained in the fibre of $\tilde{\gamma}$ through $\tilde{x}$. Because $\tilde{\gamma}$ is obviously $\pi_1(X)$-equivariant, the map $\tilde{\gamma}$ induces a fibration,called the Γ-reduction of X: $\gamma_X : X := (\widetilde{X}/\pi_1(X)) \to \Gamma(X) := \widetilde{\Gamma}(X)/\pi_1(X)$,

(This last map is constructed independently, for X projective, in [K] and termed: "Shafarevich map of X").

The geometric interpretation of γ is that if $V \subset X$ is a compact irreducible analytic subset going through a general point $x \in X$, and $\hat{V}$ its normalization, then $\pi_1(V)_X \subset \pi_1(X)$, the image of the natural map: $\pi_1(\hat{V}) \to \pi_1(X)$, is finite if and only if V is contained in the fibre of γ through x.

The fibration $f : X \to Y$ naturally induces a natural map $\gamma_f : \Gamma(X) \to \Gamma(Y)$.

4.1.1 Definition: We denote by $\gamma d(X) := dim(\Gamma(X)) \in \{0, 1, ..., n\}$, and call it the Γ-dimension of X.

Thus $\gamma d(X) = 0$ iff $\pi_1(X)$ is finite, and $\gamma d(X) = n$ iff X is of π_1-general type.

Thus $(dim(X) - \gamma d(X))$ is the maximal dimension of subvarieties of X having "finite fundamental group in X".

4.2 The invariant κ^+.

Definition: For X a compact complex connected manifold, we define : $\kappa^+(X) := max\{\kappa(X, Det(F)), F \subset \Omega^p_X, p > 0\}$, here F runs through all coherent non-zero subsheaves of $\Omega^{\cdot}_X$, $Det(F)$ is the saturation of its determinant in $\Lambda^r(\Omega^{\cdot}_X)$, and the Kodaira dimension is suitably defined (see [C95] for details).

We of course have: $\kappa^+(X) \geq \kappa(X)$, and conjecturally (see [C95]): $\kappa^+(X) = \kappa(X)$ if $\kappa(X) \geq 0$, while $\kappa^+(X) = -\infty$ iff X is rationally connected. These conjectures are consequences of the standard conjectures of MMP when X is projective, at least.

One extreme case is proved in [K], with a different formulation:

4.2.1 Proposition: *Assume that X is compact Kähler, n-dimensional, and such that: $\kappa^+(X) = n$, with K_X nef. Then: $\kappa(X) = n$.*

Proof: Let $\mathcal{F} \subset \Omega^p_X, (p > 0)$ be a coherent subsheaf of rank r , such that: $\kappa(X, \mathrm{Det}(\mathcal{F})) = n$,where $\mathrm{Det}(\mathcal{F})$ is the saturation of $\det(\mathcal{F})$ in : $\Lambda^r(\Omega^p_X)$.

By [Mi], if C is the generic curve which is the complete intersection of generic members of a sufficiently large multiple of any ample linear system, then Ω^1_X, and so also Ω^p_X restricted to C are semi-positive. The quotient (on C locally free) sheaf Q:= $\Omega^p_X/\mathcal{F}$ has thus nonnegative degree on C (ie: $\deg(Q_{/C}) \geq 0$).

Because any nef divisor is the numerical limit of ample linear systems, we obtain the inequality:

(4.2.2) $D.N_1.N_2....N_{n-1} \geq 0$ for arbitrary nef divisors $N_1, ..., N_{n-1}$,where D is the first Chern class of Q.

Now, $\mathrm{Det}(\mathcal{F})$ is big by assumption, and so there exist $\mathbb{Q}$-divisors A and E,with A ample and E effective, such that $\mathrm{Det}(\mathcal{F}) = A + E$.

For a suitable positive integer m, we have:

$m.K_X = det(\Omega^p_X) = \mathrm{Det}(\mathcal{F}) + D = A + E + D$.

And so:

$$(m.K_X)^n - A^n = (E + D) \cdot (\Sigma_{i=0}^{i=n-1} A^i.(m.K_X)^{n-1-i}) \geq 0$$

by (4.2.2).

Thus $K_X^n \geq m^{-n}.A^n > 0$, as claimed.

4.3 The comparison theorem:

It gives an inequality between κ^+ and γd.

4.3.1 Theorem ([C95]): *Let X be a compact Kähler manifold with $\chi(\mathcal{O}_X) \neq 0$. Then $\kappa^+(X) \geq \gamma d(X)$.*

Observe that if X is a complex torus, we have: $\gamma d(X) = n$, and $\kappa(X) = \kappa^+(X) = 0$. Of course, $\chi(\mathcal{O}_X) = 0$, then. It was conjectured in [C95] that X is (up to bimeromorphy) covered by a torus if $\gamma d(X) = n$, and $\kappa(X) = 0$.

Thus $\gamma d(X)$ behaves in some sense like a Kodaira dimension as well. This motivates the term "of π_1-general type", and the conjecture 0 above.

4.3.2 Corollary: *Let X be a compact Kähler n-fold. Assume that X is of π_1-general type (ie: $\gamma d(X) = n$), and that K_X is nef. Then: either $\kappa(X) = n$, or $\chi(\mathcal{O}_X) = 0$.*

Proof: Assume $\chi(\mathcal{O}_X) \neq 0$. Then $\kappa^+(X) = n$, by the comparison theorem 4.4.1. In particular, X is projective. Because K_X is nef, we obtain from 4.2.1 that $\kappa(X) = n$.

(Observe that we used the hypothesis K_X nef only in the projective case. So it is not needed to give the delicate definition in the Kähler case). Of course, we conjecture that the above result holds without the hypothesis that K_X is nef.

4.3.3 Corollary: *Let X be a compact Kähler non projective n-fold of π_1-general type. Then $\chi(\mathcal{O}_X) = 0$.*

Proof: Assume on the contrary that $\chi(\mathcal{O}_X) \neq 0$. We have: $\kappa^+(X) \geq \gamma d(X) = n$. And so X is projective. Contradiction. Thus $\chi(\mathcal{O}_X) = 0$.

4.3.4 Example: We shall consider here the very simple case of surfaces. Let X be a minimal compact Kähler surface of π_1-general type, and of non general type (ie: $\kappa(X) < 2$). We know then that X is not uniruled. If $\kappa(X) = 0$, then X is covered by a torus, by classification. If $\kappa(X) = 1$, using $\chi(\mathcal{O}_X) = 0$ (because for surfaces, $\kappa^+ = \kappa$ if $\kappa \geq 0$) and [B-P-V, III,18.2.3 and V.12.3], we see that if $f : X \to Y$ is the Iitaka-Moishezon fibration, then $f_*(\omega_{X/Y})$ is torsion, and has only fibres either smooth or multiple elliptic (of type ${}_mI_0$). And so, after a finite étale cover of X, f is a principal fibre bundle over Y with fibre some fixed elliptic curve.

§5. PROOF OF PROPOSITION 3.3.2 .

5.0 Situation. Recall that X is a compact Kähler threefold of π_1-general type, with $\kappa(X) = 1$ or 2, and that its Iitaka-Moishezon fibration $f : X \to Y$ either has base a curve Y of general type, has no multiple fibre, and has a generic fibre which is a two-dimensional complex torus, or is an elliptic fibration over a surface Y.

We want to show that X is bimeromorphic to either $Alb(X) \times_{Alb(Y)} Y$, or to a submersion with Abelian surfaces as fibres, after a suitable finite étale cover of X.

5.1 Monodromies. We shall always denote by Y^* the Zariski open subset of Y over which f is smooth, and by $\rho : \pi_1(Y^*) \to Aut(H_1(X_{y_0}, \mathbb{Z}))$ the associated monodromy representation, if $Y_0 \in Y^*$ is given. Let G be its image.

5.2 Local monodromies. If D_i is a component of the divisor $(Y - Y^*)$, and if γ_i is a small loop winding once around D_i in Y, we shall denote by T_i (as in [K,6.3]), the (well-defined up to conjugation) image of γ_i by ρ. We also denote by r_i the multiplicity of the fibre of f above a generic point of D_i.

Let $H := \pi_1(X_{y_0})_X/Torsion$ be the image in $\pi_1(X)$ of $\pi_1(X_{y_0})$, modulo its torsion subgroup. Let $q : H_1(X_{y_0}, \mathbb{Z}) \to H$ be the natural quotient map, and W its kernel.

Easy arguments (see [K,6.3] for details) show that $N := \sum_{i \in I} Ker(1 - T_i^{r_i}) \subset W$

We write: $N_{\mathbb{Q}} := N \otimes_{\mathbb{Z}} \mathbb{Q}$.

5.3 Proposition: *For f as in 5.0 above, $N_{\mathbb{Q}} \subset H_1(X_{y_0}, \mathbb{Q})$ is a sub Hodge structure.*

(This means here that the orthogonal of N in $H^1(X_{y_0}, \mathbb{R})$ is a complex subspace, when $H^1(X_{y_0}, \mathbb{R})$ is identified with $H^0(X_{y_0}, \Omega^1_{X_{y_0}})$ in the Hodge decomposition of $H^1(X_{y_0}, \mathbb{C})$).

This holds in a much broader context, in fact: all Kähler fibrations, but the proof seems to be written only when the fibres are projective varieties, so we must appeal to other arguments when the fibres are not algebraic.

Proof: It rests on results stated below, but whose proofs do not depend on the statement 5.3. We distinguish three cases:

1. $\kappa(X) = 1$, and X_y is not projective. Then f has "constant moduli" (ie: 5.5 holds). The monodromy is then (dually) a finite subgroup of $Aut_{\mathbb{C}}(H^0(X_{y_0}, \Omega^1))$, this vector space being identified with $(H^1(X_{y_0}, \mathbb{R}))$ by means of the Hodge decomposition on X_{y_0}. See the proof of 5.6 for details.

2. $\kappa(X) = 2$. The fibres are projective, here. Hence the results of Deligne [D] apply and give the claim. We give another, more elementary argument. Let $C \subset Y$ be a generic very ample curve. Let $X_C = f^*(C)$. Then X_C is an elliptic surface of π_1-general type, with $\kappa = 1$. From Example 4.3.4, we learn that after a finite étale cover, this surface is a principal fibre bundle with fibre an elliptic curve. The monodromy is thus finite, and complex linear in this case too (because $\pi_1(C \cap Y^*)$ maps surjectively to $\pi_1(Y^*)$, by Lefschetz theorem, and monodromy is compatible with restrictions).

3. $\kappa(X) = 1$, and X_y is projective. In this case, the fibres of f are projective, and so the results of [D] apply. Alternatively, we can use the untwisted model of f (see 5.5.1 below), which is projective, and has the same monodromy as f to apply the arguments of [K,6.3]. .

5.4 Corollary: *Let $f : X \to Y$ be as in 5.0. Then $T_i = 1, \forall i \in I$.*

Proof: If $N \neq 0$, then there is a non-zero sub Hodge structure of $H^1(X_{y_0}, \mathbb{Q})$ mapped to 1 in $\pi_1(X)$. But this sub Hodge structure defines a non-zero complex subtorus A of X_{y_0}, whose π_1 is mapped to 1 in $\pi_1(X)$. This is a contradiction to the fact that X is of π_1-general type.

Thus: $(1 - T_i^{r_i}), \forall i \in I$. If $\kappa(X) = 1$, we have $r_i = 1, \forall i$, because f has no multiple fibres. And so 5.4 holds in this case.

Now, if $\kappa(X) = 2$, we saw in the proof of 5.3 that all multiple fibres of f are of type ${}_mI_0$, which directly implies that $T_i = 1$.

5.5 Lemma: *Let $f : X \to Y$ be as in 5.0. There is a fixed complex torus F such that $X_y \cong F$, for any smooth fibre X_y of f, unless maybe when $\kappa(X) = 1$ and the generic fibres X_y of f are Abelian surfaces. In this case, f is bimeromorphic to a submersion with Abelian surfaces as fibres.*

Proof: When $\kappa(X) = 2$, this is an immediate consequence of example 4.3.4, and the proof of 5.3 in this case, by taking the surface $X_C = f^{-1}(C)$, for C a generic very ample curve of Y.

When $\kappa(X) = 1$, and the general fibre X_y of f is a non-projective torus, this follows from [F], or [CP, 7.1]. Note that this is stated only for f the algebraic reduction of X there, but that the proof actually works for any fibration.

The remaining case is when $\kappa(X) = 1$, and X_y is an abelian surface.

For this, we need to consider here:

5.5.1 The untwisted Model of f. For any fibration $f : X \to Y$, with X compact Kähler, and with generic fibres X_y complex tori, we introduced in [CP,6.3] the so-called "untwisted" model $f' : X' \to Y'$ of f. This fibration is caracterised up to bimeromorphic equivalence by the following two properties:

1. For $y \in Y$ generic, we have: $X_y \cong X'_y$.
2. f' has a meromorphic section.

The construction is sketched in [CP, 6.3]. It was used already in [F] and [C85].

Recall from [CP,6.5] that $h^p(X, \mathcal{O}_X) = h^p(X', \mathcal{O}_{X'}), \forall p \geq 0$.

In particular, $q(X) = q(X')$.

5.5.2 Comparison of monodromies. Let now, as always, denote by Y^* the Zariski open subset of Y over which f and f' are smooth, and by $\rho : \pi_1(Y^*) \to Aut(H_1(X_{y_0}, \mathbb{Z})$ and $\rho' : \pi_1(Y^*) \to Aut(H_1(X'_{y_0}, \mathbb{Z})$ the associated monodromy representations, if $Y_0 \in Y^*$ is given. We trivially have (identifying X_{y_0} and X'_{y_0}):

5.5.3 Lemma: $\rho = \rho'$.

Proof: The fibration f' is obtained from f over Y^* simply by twisting locally by means of translations in the fibres (as seen from the Chow Scheme construction of f'). These translations act trivially on the cohomology.

We can now conclude the proof of 5.5 in the case considered ($\kappa(X) = 1$, and X_y is an abelian surface):

First: because X_y is projective for $y \in Y$ generic, it follows from [C81] that X' is projective (although X is not), because f' has a section. Next, we have seen that the local monodromies of f are trivial. Thus the same holds for the local monodromies of f'. Because f' has a section, we learn from [K,6.8] that after a base change over a finite étale cover of Y, X' is bimeromorphic to an abelian group scheme over Y, corresponding to a map $\mu : Y \to \mathcal{A}_3$, the moduli of polarised Abelian surfaces with a level-three structure.

Because the fibration f has no multiple fibres, and its generic fibres are complex tori, it has local sections (in the analytic topology) over its one-dimensional base. Thus f and f' are locally (over Y) bimeromorphic. Hence the claim.

5.6 Lemma: *Let $f : X \to Y$ be as in 5.0. The monodromy becomes trivial, after a finite étale cover of X, and $q(X) = q(Y) + q(F)$, unless maybe when $\kappa(X) = 1$ and the generic fibre X_y of f is an Abelian surface.*

Proof: The second assertion follows from the first and 5.5, using [B]. For the first assertion, we show that the monodromy representation ρ has finite image contained (dually) in the complex linear automorphisms of $H^1(X_{y_0}, \mathbb{R})$. This is an easy consequence of the compactness of the Chow Scheme of X in the Kähler case: just consider indeed the connected component $\mathcal{C}$ of the Chow Scheme of $(X \times_Y X)/Y$ which for y generic in Y consists of isomorphisms of F with X_y (F a fixed torus isomorphic to X_y, the existence of which is the content of 5.5), this component containing a given isomorphism of F with X_{y_0}. The connected components of the fibre of $\mathcal{C}$ over y_0 then consists of exactly the isomorphisms of F with X_{y_0} obtained by composing the given one with the image of ρ, and with translations of X_{y_0}. This shows the assertion.

We finally obtain 3.3.1 as follows: we saw that either $\kappa(X) = 1$, and f is bimeromorphically a submersion onto a curve of general type with Abelian surfaces as fibres, or we reduced to the case where: the global monodromy of f is trivial, the generic fibre is a fixed torus F, and $q(X) = q(Y) + q(F)$. This last case we now treat.

We consider the two cases $\kappa(X) = 1, 2$ separately again:

If $\kappa(X) = 1$, then f has no multiple fibres and trivial monodromy, it is then easy to check that the natural map: $v : X \to V := Alb(X) \times_{Alb(Y)} Y$ is bimeromorphic (the details for the case $\kappa(X) = 2$ below apply to show this).

If $\kappa(X) = 2$, we have to eliminate the multiple fibres of f by a finite étale cover of X. In this case, the natural map $v : X \to Alb(X) \times_{Alb(Y)} Y$ is surjective and generically finite (by 5.6).

Moreover, its restriction to a generic fibre X_y of f is étale, and maps X_y to the corresponding fibre V_y of V over Y. The natural map: $Alb(v) : Alb(X) \to Alb(V)$ is isomorphic, by construction, but $\pi_1(Alb(X_y))$ maps to a subgroup K' of finite index in $K := Ker(\pi_1(Alb(X)) \to \pi_1(Alb(Y))$. Choose a subgroup of finite index A' in $\pi_1(Alb(X))$ such that $A' \cap K = K'$, and let the corresponding étale cover of $Alb(X)$ be $Alb'(X)$. Replacing X by $X' := Alb'(X) \times_{Alb(X)} X$ then gives the desired étale cover of X. This is because the natural map from X' to $V' := Alb'(X) \times_{Alb(Y')} Y$, (where Y' is the Stein factorisation of the composite map $X' \to Y$) maps bijectively $X'_{y'}$ to $V'_{y'}$.

§6. PROOF OF COROLLARY 2 OF THEOREM 1.

We treat the situation case-by-case, assuming $\kappa(X) < 3$. We know from theorem 1 that $\kappa(X) \geq 0$, and that $X = Alb(X) \times_{Alb(Y)} Y$, if $f : X \to Y$ is the Iitaka-Moishezon fibration.

If $\kappa(X) = 0$, then Y is a point, and $X = Alb(X)$.

If $\kappa(X) = 1$, then Y is a curve of general type (by 3.3.1) isomorphic to $Alb(Y)$. Thus either $X = \alpha_X(X) \subset Alb(X)$, the Albanese image of X, or X has a submersion onto the curve Y of general type with fibres Abelian surfaces (cases 2.a, 2.b respectively).

If $\kappa(X) = 2$, we shall distinguish three cases, according to $d : dim(\alpha_Y(Y))$, the dimension of the Albanese image of Y.

3.a occurs of course exactly when $d = 2$, so that α_Y is generically finite.

3.b occurs when $d = 0$, for any finite étale cover of Y.

3.c occurs when $d = 1$, so that C is the Albanese image of Y. (Notice that C is indeed of general case, after a suitable finite étale cover of Y, because $\kappa(Y) = 2$).

References

[B-P-V] W.BARTH, C.PETERS, A.VAN DE VEN. *Compact Complex Surfaces.* Springer-Verlag 1984.

[B] A.BLANCHARD. *Sur les variétés analytiques complexes.* Ann.Sc.ENS 73 (1956), 157-202.

[81] F.CAMPANA. *Coréduction algébrique d'un espace aalytique faiblement Kählérien compact.* Inv. Math. (1981), 187-223.

[C85] F. CAMPANA *Réduction d'Albanese d'un morphisme Kählérien propre et applications*, I et II.Comp. Math.54 (1985), 373-398 et 399-416.

[C95] F. CAMPANA *Fundamental group and positivity properties of cotangent bundles of compact Kähler manifolds.* J. Alg. Geom., **4**, p. 487-502, (1995).

[C95'] F. CAMPANA *Remarques sur les groupes de Kähler nilpotents.* Ann. ENS, **28**, p. 307-316, (1994).

[C98'] F. CAMPANA *Negativity properties of compact curves in infinite covers of projective surfaces.* J. Alg. Geom.**7**, p. 673-693, (1998).

[C01] F. CAMPANA *Green-Lazarsfeld sets and solvable quotients of Kähler groups.* (10) J. Alg. Geom. (2001), 599-622.

[CP] F.CAMPANA, T.PETERNELL. *Complex Threefolds with non-trivial Holomorphic 2-Forms.* J.Alg.Geom. **9**(2000), 223-264.

[D] P.DELIGNE. *Theorie de Hodge II.* Publ.Math.IHES **40**,5-57 (1972)

[I] S.IITAKA *Genera and classification of Algebraic Varieties.* Sugaku 24 (1972), 14-27.

[Kaw] Y. KAWAMATA *Characterization of Abelian Varieties.* Comp. Math., **43**, p. 253-276, (1981).

[K] J.KOLLÁR *Shafarevich maps and plurigenera of Algebraic Varieties.* Inv.Math., **113**, p. 177-215,(1993).

[K-O] S.KOBAYASHI, T.OCHIAI. *Meromorphic Mappings into compact complex spaces of general type.* Inv. Math. 31 (1975),7-16.

[Mi] Y.MIYAOKA *On the Kodaira Dimension of a Minimal Threefold.* Math. Ann. 281 (1988), 325-332.

[Moi] B.MOISHEZON *Algebraic Varieties and Compact Complex Spaces.* Actes du Congrès International des Mathématiciens. Nice 1970. Vol. 2, 643-648.

[Mo] S.MORI *Flip Theorem and the existence of minimal models for threefolds.* J.AMS. 1 (1988),117-253.

[N'] N.NAKAYAMA. *Projective algebraic varieties whose universal covering spaces are biholomorphic to $\mathbb{C}^n$.* J. Math. Soc. Japan 51 (1999),645-654.

[N"] N.NAKAYAMA. *Compact Kähler manifolds whose universal covering spaces are biholomorphic to $\mathbb{C}^n$.* RIMS preprint-1230(1999),1-100.

[N-S] Y.NAMIKAWA, JHM. Steenbrink. *Global smoothing of Calabi-Yau Threefolds.* Invent. Math. 122 (1995), 403-419.

[U] K. UENO *Classification theory of Algebraic varieties* LNM, **439**, Springer-Verlag (1975).

F. Campana

Département De Mathématiques

Université Nancy 1

P 239, F-54506 Vandoeuvre-Nancy Cedex

France

Email: campana@iecn.u-nancy.fr

Q. Zhang
Department of Mathematics
University of Missouri
Columbia, MO 65211
U.S.A.
Email: qi@math.missouri.edu

Contemporary Mathematics
Volume **386**, 2005

COMPLETIONS OF VALUATION RINGS

STEVEN DALE CUTKOSKY AND LAURA GHEZZI

1. Introduction

Suppose that k is a field of characteristic zero, K is an algebraic function field over k, and V is a k-valuation ring of K (that is, $k \subset V$ and the quotient field of V is K). Zariski's theorem of local uniformization [23] shows that there exist algebraic regular local rings R_i with quotient field K which are dominated by V, and such that the direct limit

$$\cup R_i = V.$$

Now suppose that K^* is a finite algebraic extension of K and V^* is a k-valuation ring of K^* such that $V = V^* \cap K$. Let Γ^* be the value group of V^*, and Γ be the value group of V.

The first author has shown with Olivier Piltant in [9] that a relative local uniformization theorem holds for the extension K^* of K, which gives the strongest possible generalization of the classical ramification theory of Dedekind domains to general valuations. The following theorem is a summary of some of the conclusions of Theorem 6.3 [9].

Theorem 1.1. *Suppose that assumptions and notations are as above. Let k' be an algebraic closure of V^*/m_{V^*}. Then there exist a directed system of algebraic regular local rings S_i with quotient field K^* which are dominated by V^*, and a directed system of algebraic normal local rings R_i with quotient field K which are dominated by V such that*

(1) $\cup S_i = V^*$ *and* $\cup R_i = V$.
(2) S_i *is a localization at a maximal ideal of the integral closure of R_i in K^* for all i.*
(3) *There exist actions of Γ^*/Γ on $\hat{S}_i \hat{\otimes}_{S_i/m_{S_i}} k'$ which are compatible with the directed system so that*

$$(\hat{S}_i \hat{\otimes}_{S_i/m_{S_i}} k')^{\Gamma^*/\Gamma} \cong \hat{R}_i \hat{\otimes}_{R_i/m_{R_i}} k'.$$

It was shown by an example of Abhyankar [2] that it is in general not possible to find an algebraic regular local ring S with quotient field K^* which is dominated by V^* such that there exists an algebraic regular local ring R with quotient field K such that S is a localization of the integral closure of R in K^*. The fact (proven in [9] and [7]) that normal local rings R exist satisfying this property proves the "local weak simultaneous resolution conjecture" of Abhyankar, posed by Abhyankar in [2] and [5]. The R_i found in the proof of Theorem 1.1 in fact have toric singularities. This is reflected in the fact stated above that their completions are abelian quotient singularities.

From this theorem we obtain the following.

Research of the first author was partially supported by NSF.

Theorem 1.2. *Let notations be as in Theorem 1.1. Let* $U^* = \cup \hat{S}_i \hat{\otimes}_{S_i/m_{S_i}} k'$ *and let* $U = \cup \hat{R}_i \hat{\otimes}_{R_i/m_{R_i}} k'$*. Then* U^* *and* U *are Henselian normal domains, and* $Q(U^*)$ *is a finite Galois extension of* $Q(U)$ *with Galois group* Γ^*/Γ.

We give a proof of Theorem 1.2 in Section 3.

In this paper we compare the "completion" of Theorem 1.2 with other notions of completion of a valuation ring ([16], [18], [19], [14], [21], [22]).

Let us briefly allow k to be an arbitrary field. We summarize some of the results of Section 4. Suppose that $\{R_i\}$ is a directed system of normal algebraic local rings which are dominated by V, and such that $\cup R_i = V$. The ring $T = \cup \hat{R}_i$ does not depend on our choice of $\{R_i\}$ whose union is V (Lemma 4.1), and is Henselian (Proposition 4.2). Thus T can be considered to be a "completion" of the valuation ring V. We give an example showing that T is in general not a valuation ring, and we show that T is itself a valuation ring if and only if for each i there exists a unique valuation ring V_i with quotient field K_i (where K_i is the quotient field of $\hat{R}_i$) which dominates V and $\hat{R}_i$. (Theorem 4.4). We make use of a theorem of Heinzer and Sally [14] on the uniqueness of extensions of valuations dominating a local ring to their completion in proving this result.

We give an example (Example 7.5) showing that even if V and T are rank 1 valuation rings, then T is in general not complete and in particular is not a maximal immediate extension, as defined in [18] and [16].

The essential obstruction to T being a valuation ring is the problem of the rank of the valuation increasing upon extending the valuation dominating a particular R to a valuation dominating its completion (Corollary to Theorem 4.4). In the case of rank 1 valuations, this problem can be handled in a very satisfactory way, and (in characteristic zero) we will obtain a good valuation theoretic explanation of Theorem 1.2.

In Section 5, we define the prime ideal $p(\hat{R})_\infty$ of elements of infinite value of the completion $\hat{R}$ of an algebraic local ring R dominated by a rank 1 valuation V. This prime has previously been defined and considered in [6] and [22], as well as by Spivakovsky. The essential point here is that there is a unique extension of the valuation ring V to a valuation ring of the quotient field of $\hat{R}/p(\hat{R})_\infty$ which dominates $\hat{R}/p(\hat{R})_\infty$. We conclude that there is a unique valuation ring $\overline{V}$ of the quotient field of the ring $\overline{T} = \cup \hat{R}_i/p(\hat{R}_i)_\infty$ which contains T.

In the case when V has rank greater than 1 there is no natural ideal in $\hat{R}$ which contains the obstruction to the jumping of the rank of an extension of V to $\hat{R}$, although this obstruction is obtained in a series of prime ideals in quotient rings of $\hat{R}$.

For the remainder of this introduction we assume that k has characteristic zero, and V has rank 1. We prove that $\overline{T}$ is in fact a valuation ring in Theorem 7.3, and that $(\overline{T}, Q(\overline{T}))$ is an immediate Henselian extension of (V, K) in Theorem 7.4.

We further show (in Theorem 7.4) that we can choose our system of regular local rings R_i so that each $\hat{R}_i/p(\hat{R}_i)_\infty$ is a regular local ring. The main new technical result used in this statement is Theorem 6.5, which shows that we can simultaneously resolve the primes of infinite value in a finite extension. In this case the "finite extension" is just the identity, but we will need this more general result later.

We now turn to an analysis of our finite extension K^* over K, in the case when V^* (and $V = V^* \cap K$) are rank 1 valuation rings and k has characteristic zero. We make essential use of Theorem 6.5 (on simultaneous resolution of the primes of infinite value). We obtain in Theorems 8.1 and 8.2 a generalization of Theorem 5.1 [6] and Theorem 1.1 (Theorem 6.3 [9]) in this context. Let k' be an algebraic closure of

V^*/m_{V^*}. We find a system of regular local rings S_i whose union is V^*, and a system of normal local rings R_i whose union is V, such that for all i, S_i is a localization at a maximal ideal of the integral closure of R_i in K^*. If q_i and p_i are the respective primes of infinite value, then $\hat{S}_i/q_i$ is a regular local ring and $\hat{R}_i/p_i$ is a normal local ring with toric singularities. There are compatible actions of Γ^*/Γ on $(\hat{S}_i/q_i)\hat{\otimes}_{S_i/m_{S_i}}k'$ such that

$$((\hat{S}_i/q_i)\hat{\otimes}_{S_i/m_{S_i}}k')^{\Gamma^*/\Gamma} \cong (\hat{R}_i/p_i)\hat{\otimes}_{R_i/m_{R_i}}k'.$$

From Theorems 8.1 and 8.2 we obtain the following.

Theorem 1.3. *Let notations be as in Theorem 1.1. Assume that V^* (and $V = V^* \cap K$) have rank 1, and that $k = V^*/m_{V^*}$ is algebraically closed of characteristic zero. Then there exist directed systems of algebraic local rings R_i and S_i satisfying the conclusions of Theorem 1.1 and such that $\hat{R}_i/p_i$ and $\hat{S}_i/q_i$, where q_i and p_i are the primes of elements of infinite value, satisfy the conclusions of the above paragraph. Let $\overline{U}^* = \cup \hat{S}_i/q_i$ and let $\overline{U} = \cup \hat{R}_i/p_i$. Then $\overline{U}^*$ and $\overline{U}$ are Henselian valuation rings, such that $(\overline{U}^*, Q(\overline{U}^*))$ and $(\overline{U}, Q(\overline{U}))$ are immediate extensions of (V^*, K^*) and (V, K) respectively, and $Q(\overline{U}^*)$ is a finite Galois extension of $Q(\overline{U})$ with Galois group Γ^*/Γ.*

We give the proof of Theorem 1.3 in Section 8.

2. Notations

We will denote the maximal ideal of a local ring R by m_R or $m(R)$. We will denote the quotient field of a domain R by $Q(R)$. Suppose that $R \subset S$ is an inclusion of local rings. We will say that R dominates S if $m_S \cap R = m_R$. Suppose that K is an algebraic function field over a field k. We will say that a subring R of K is algebraic if R is essentially of finite type over k. Suppose that K^* is a finite extension of an algebraic function field K, R is a local ring with quotient field K and S is a local ring with quotient field K^*. We will say that S lies over R and R lies below S if S is a localization at a maximal ideal of the integral closure of R in K^*. If R is a local ring, $\hat{R}$ will denote the completion of R at its maximal ideal.

Good introductions to the valuation theory which we require in this paper can be found in Chapter VI of [24] and in [3]. A valuation ν of K will be called a k-valuation if $\nu(k) = 0$. We will denote by V_ν the associated valuation ring, which necessarily contains k. A valuation ring V of K will be called a k-valuation ring if $k \subset V$. The residue field V/m_V of a valuation ring V will be denoted by $k(\nu)$. The value group of a valuation ν with valuation ring V will be denoted by Γ_ν or Γ_V. If R is a subring of V_ν then the center of ν (the center of V_ν) on R is the prime ideal $R \cap m_{V_\nu}$.

Suppose that R is a local domain. A monoidal transform $R \to R_1$ is a birational extension of local domains such that $R_1 = R[\frac{P}{x}]_m$ where P is a regular prime ideal of R, $0 \neq x \in P$ and m is a prime ideal of $R[\frac{P}{x}]$ such that $m \cap R = m_R$. $R \to R_1$ is called a quadratic transform if $P = m_R$.

If R is regular, and $R \to R_1$ is a monoidal transform, then there exists a regular system of parameters $(x_1, \ldots, x_n)$ in R and $r \leq n$ such that

$$R_1 = R\left[\frac{x_2}{x_1}, \ldots, \frac{x_r}{x_1}\right]_m.$$

Suppose that ν is a valuation of the quotient field R with valuation ring V_ν which dominates R. Then $R \to R_1$ is a monoidal transform along ν (along V_ν) if ν dominates R_1.

3. Completion of relative local uniformization

We now give the proof of Theorem 1.2.

Let $\overline{S}_i = \hat{S}_i\hat{\otimes}_{S_i/m_{S_i}}k'$, $\overline{R}_i = \hat{R}_i\hat{\otimes}_{R_i/m_{R_i}}k'$ for $i \in I$ and let $G = \Gamma^*/\Gamma$. U^* is the directed union of the $\overline{S}_i$ and U is the directed union of the $\overline{R}_i$. The fact that R_i is a normal domain implies $\hat{R}_i$ is a normal domain (Chapter VIII, Section 13, Theorem 32 [24]) and thus $\overline{R}_i$ is a normal local domain (Proposition IV.6.7.4 [11]). Thus U (and U^*) are normal domains. The fact that U and U^* are Henselian follows from the proof of Proposition 4.2. The action of G on U^* extends to an action on $Q(U^*)$. Suppose that $h \in U^*$ and $\sigma(h) = h$ for all $\sigma \in G$. There exists i such that $h \in Q(\overline{S}_i)$. Since $\overline{S}_i$ is finite over $\overline{R}_i$ and $\overline{S}_i^G = \overline{R}_i$, it follows that $Q(\overline{S}_i)^G = Q(\overline{R}_i)$. Thus $h \in Q(U)$. We conclude that $Q(U^*)^G = Q(U)$, so that $Q(U^*)$ is a finite Galois extension of $Q(U)$ (c.f. Theorem V.2.15 [15]).

Remark 3.1. *The statement that $Q(U^*)$ is finite over $Q(U)$ can be seen directly from the fact that the minimal polynomial of each $Q(\hat{S}_i)$ over $Q(\hat{R}_i)$ is a factor of the minimal polynomial of an appropriate primitive element of K^* over K (by Proposition 1 [1]).*

The fact that the extension considered in Theorem 1.2 is Galois, even when the original field extension K^*/K is not, is a condition that can be easily seen in the case when U and U^* are valuation rings, as the first author realized with Franz-Viktor Kuhlmann in a discussion.

Theorem 3.2. *Suppose that V is a Henselian valuation ring of a field K, such that V contains an algebraically closed field k of characteristic zero, with $k \cong V/m_V$. If L is a finite extension of K, then there is a unique valuation ring W of L such that W dominates V, and L is Galois over K with Galois group Γ_W/Γ_V, where Γ_W and Γ_V are the respective value groups.*

Proof. Let J be a finite Galois extension of K which contains L. Let G be the Galois group of J over K. Since V is Henselian, there exists a unique valuation ring U of J such that U dominates V ((16.4), (16.6) [10]). Thus the splitting group $G^s(U/V) = G$ by Proposition 1.46 [3]. We have $U/m_U = V/m_V = k$ since k is algebraically closed. Thus the inertia group $G^i(U/V) = G^s(U/V) = G$ by Theorem 1.48 [3]. Finally,

$$G = G^i(U/V) \cong \Gamma_U/\Gamma_V$$

by Theorem 3 [17] or Chapter VI, Section 12, Corollary [24].

Since G is abelian, all intermediate subfields of J are Galois over K. Thus L is Galois over K, and the Galois group of L over K is Γ_W/Γ_V. □

4. Completions of valuation rings

Suppose that K is an algebraic function field over a field k, and V is a valuation ring of K with maximal ideal m_V and value group Γ. Suppose that $\{R_i \mid i \in I\}$ is a directed system of normal local rings such that

(a) $V = \cup_{i\in I}R_i$.
(b) I has a minimum 0.
(c) Each R_i is essentially of finite type over k and has quotient field K.
(d) If $i < j$ then R_j dominates R_i.

Let K_i be the quotient field of $\hat{R}_i$. By Zariski's subspace theorem ((10.13) [4]) we have natural inclusions $\hat{R}_i \to \hat{R}_j$ if $i < j$, and $\{\hat{R}_i \mid i \in I\}$ is a directed system of normal local rings (Scholie 7.8.3 [11]). Let

$$T = \cup_{i\in I}\hat{R}_i$$

and $K_\infty = \cup K_i$. T is a normal domain with quotient field K_∞ and maximal ideal $m_T = \cup m_{\hat{R}_i}$.

Given a valuation ring V as above, there exists a directed system of normal local rings $\{R_i\}$ whose union is V. A particular construction is as follows. We take R_0 to be any normal local ring which is dominated by V. If $m \in \mathbf{N}$ and $f_1, \ldots, f_m \in V$ we set $i = (f_1, \ldots, f_m)$ and let R_i be the localization of the normalization of $R_0[f_1, \ldots, f_m]$ which is dominated by V.

Lemma 4.1. *The ring $T = \cup_{i\in I}\hat{R}_i$ is independent of choice of directed system $\{R_i \mid i \in I\}$ satisfying (a),(b),(c), and (d).*

Proof. Let J be a partially ordered set, and let $\{S_j \mid j \in J\}$ be a collection of algebraic local rings with quotient field K, such that $\{S_j\}$ satisfies (a),(b),(c) and (d). We show that $\cup_{i\in I}\hat{R}_i = \cup_{j\in J}\hat{S}_j$.

Let $i \in I$. Since R_i is essentially of finite type over k and dominated by V, there exist $f_1, \ldots, f_m \in V$ such that $R_i = k[f_1, \ldots, f_m]_{m_V \cap k[f_1,\ldots,f_m]}$. Since $V = \cup_{j\in J} S_j$ and J is directed, there exists $j \in J$ such that $f_1, \ldots, f_m \in S_j$, and so $k[f_1, \ldots, f_m]_{m_V \cap k[f_1,\ldots,f_m]} \subset S_j$, since S_j is dominated by V. Hence $R_i \subset S_j$. There is then a natural inclusion $\hat{R}_i \subset \hat{S}_j$, and thus $\cup_{i\in I}\hat{R}_i \subset \cup_{j\in J}\hat{S}_j$. The other inclusion is proven in the same way. □

Proposition 4.2. *The ring $T = \cup_{i\in I}\hat{R}_i$ is Henselian.*

Proof. Let $F \in T[x]$ and $\phi_1, \phi_2 \in T/m_T[x]$ be monic polynomials such that ϕ_1 and ϕ_2 are relatively prime and $\bar{F} = \pi(F) = \phi_1\phi_2$, where $\pi : T \to T/m_T$ is the natural projection. We need to show that there exist monic polynomials $F_1, F_2 \in T[x]$ such that $\pi(F_1) = \phi_1$, $\pi(F_2) = \phi_2$, and $F = F_1F_2$.

Since both $T = \cup\hat{R}_i$ and $T/m_T = \cup\hat{R}_i/m_{\hat{R}_i}$ are directed unions, there exists $c \in I$ such that $F \in \hat{R}_c[x]$ and $\phi_1, \phi_2 \in \hat{R}_c/m_{\hat{R}_c}$.

Since $\hat{R}_c$ is complete, there exist monic polynomials $F_1, F_2 \in \hat{R}_c[x]$ such that $\pi'(F_1) = \phi_1$, $\pi'(F_2) = \phi_2$, and $F = F_1F_2$. Since $F_1, F_2 \in T[x]$, F_1 and F_2 are the desired polynomials. □

Example 4.3. *In general, T is not a valuation ring.*

Proof. Let k be a field and $R = k[x, y, z]_{(x,y,z)}$. We will define a valuation ν on $K = Q(R)$ which dominates R. Let $p(t) = \sum_{i=1}^{\infty} a_i t^i \in k[[t]]$ be a transcendental series. If $f \in k[x, y, z]$ write $f = z^r g(x, y, z)$ where $g(x, y, 0) \neq 0$. We define

$$\nu(f) = (r, \text{ ord } g(t, p(t), 0)).$$

The value group of ν is $\mathbf{Z} \oplus \mathbf{Z}$ with the lexicographic order. Let V be the valuation ring of ν and let R_i be a directed system of regular local rings satisfying (a), (b), (c) and (d) with $R_0 = R$. Such a system exists by Theorem 1.1. In particular, $V = \cup R_i$.

We will suppose that T is a valuation ring and derive a contradiction. Let $\overline{\nu}$ be an extension of ν to $Q(T)$ such that T is the valuation ring of $\overline{\nu}$. There is a natural embedding of value groups

$$\Gamma_\nu = \mathbf{Z} \oplus \mathbf{Z} \subset \Gamma_{\overline{\nu}}.$$

$\frac{z}{y-p(x)} \in Q(T)$. If $\overline{\nu}(\frac{z}{y-p(x)}) \geq 0$ let $f = \frac{z}{y-p(x)}$. If $\overline{\nu}(\frac{z}{y-p(x)}) < 0$ let $f = \frac{y-p(x)}{z}$. By construction $\overline{\nu}(f) \geq 0$ so that $f \in T$ (by our assumption that T is a valuation ring). Thus there exists $\hat{R}_i$ such that $f \in \hat{R}_i$. By our hypothesis, R_i dominates R. Since

R_i is essentially of finite type over R, $Q(R) = Q(R_i)$ and R is a UFD, there exists an ideal $I \subset R$ of height ≥ 2 and $a \in I$ with

$$\nu(a) = \min\{\nu(h) \mid h \in I\}$$

such that $R[\frac{I}{a}] \subset V$ and

$$R_i = R[\frac{I}{a}]_{m_V \cap R[\frac{I}{a}]}$$

Since $I \not\subset (z)$, we have that $\nu(a) \in \{0\} \oplus \mathbf{Z}$. Let $R' = \hat{R}[\frac{I}{a}]_{m_T \cap \hat{R}[\frac{I}{a}]}$.

$$f \in Q(\hat{R}) \cap \hat{R}_i = Q(R') \cap \hat{R}' = R'.$$

For the last equality, c.f. Lemma 2.1 [6].

We can thus write $f = \frac{g}{h}$ with $g, h \in \hat{R}[\frac{I}{a}]$ and $h \notin m_T \cap \hat{R}[\frac{I}{a}]$. Thus $\overline{\nu}(h) = 0$. There exists $n \in \mathbf{N}$ such that $a^n g = g_0$, $a^n h = h_0$ and $g_0, h_0 \in \hat{R}$. Thus $\overline{\nu}(h_0) = n\nu(a) + \overline{\nu}(h) = (0, m)$ for some $m \in \mathbf{N}$.

If $f = \frac{z}{y-p(x)}$, we have $zh_0 = (y - p(x))g_0$ implies $y - p(x)$ divides h_0 in $\hat{R}$. But

$$y - p(x) = (y - \sum_{i=1}^{m+1} a_i x^i) - x^{m+2}(\sum_{i=m+2}^{\infty} a_i x^{i-m-2}).$$

Thus $\overline{\nu}(y - p(x)) > (0, m)$, which is a contradiction.

If $f = \frac{y-p(x)}{z}$, then $(y - p(x))h_0 = zg_0$ implies z divides h_0 in $\hat{R}$.
This is a contradiction since $\nu(z) = (1, 0) > (0, m) = \overline{\nu}(h_0)$.

□

As an extension of the above example, we construct valuations ν_1, ν_2 of $Q(\hat{R})$ which extend ν and dominate $\hat{R}$ such that ν_1 is an immediate extensions of ν ($k(\nu) = k(\nu_1)$ and $\Gamma_\nu = \Gamma_{\nu_1}$) but ν_2 is a rank 3 valuation, of higher rank than ν.

We first define ν_1. For $f \in \hat{R}$, write

$$f(x, y, z) = \sum a_{ij}(y - p(x))^i z^j \tag{1}$$

with $a_{ij} \in k[[x]]$. Let

$$a = \min\{i + j \mid a_{ij}(x) \neq 0\},$$
$$b = \min\{\text{ord}(a_{ij}(x)) \mid i + j = a\}.$$

Set $\nu_1(f) = (a, b) \in \mathbf{Z} \oplus \mathbf{Z}$. ν_1 defines a valuation on $Q(\hat{R})$ which dominates $\hat{R}$. We will now verify that ν_1 extends ν. For $f \in k[x, y]$, write

$$f = \sum b_j(x, y) z^j$$

with each $b_j \in k[x, y]$. Suppose that $\nu(f) = (r, s)$. Comparing with (1), we see that for all j,

$$b_j(x, y) = \sum_{i=0}^{\infty} a_{ij}(x)(y - p(x))^i.$$

Thus we either have that $b_j = 0$ or $a_{0j} \neq 0$, since $y - p(x) \not| b_j(x, y)$. Thus

$$a = \min\{j \mid b_j(x, y) \neq 0\} = r.$$

Set $g = \frac{f}{z^r}$. We have

$$b = \text{ord}(a_{0r}(x)) = \text{ord}(g(x, p(x))) = s.$$

Thus $\nu_1(f) = \nu(f)$. It follows that ν_1 is an immediate extension of ν.

Now we define the extension ν_2. For $f \in \hat{R}$, write $f = z^\alpha g(x, y, z)$ where $z \not| g(x, y, z)$. Write $g(x, y, 0) = (y - p(x))^\beta h(x, y)$ where $y - p(x) \not| g(x, y, 0)$. Set

$\gamma = \text{ord}(h(x, p(x)))$. Define $\nu_2(f) = (\alpha, \beta, \gamma) \in \mathbf{Z}^3$, where $\mathbf{Z}^3$ has the lexicographic order. ν_2 extends to a valuation of $Q(\hat{R})$ which dominates $\hat{R}$, and such that ν_2 extends ν.

Theorem 4.4. *T is a valuation ring if and only if for all i, there exists a unique valuation ring V_i with quotient field K_i which dominates V and $\hat{R}_i$.*

Proof. Suppose that for all i, there exists a unique valuation ring V_i with quotient field K_i which dominates V and $\hat{R}_i$. By Proposition 4.1 [14]

$$V_i = \hat{R}_i[V]_{m_V \hat{R}_i[V]} \subset K_i$$

is a valuation ring for all i. Let

$$V_\infty = \cup V_i = (\cup \hat{R}_i)[V]_{m_V (\cup \hat{R}_i)[V]} \subset K_\infty$$

where K_∞ is the quotient field of T. $f \in K_\infty$ implies $f \in K_i$ for some i which implies $f \in V_i$ or $\frac{1}{f} \in V_i$. Thus $f \in V_\infty$ or $\frac{1}{f} \in V_\infty$. Thus V_∞ is a valuation ring. $V \subset \cup \hat{R}_i$ implies $V_\infty = \cup \hat{R}_i = T$, and we conclude that T is a valuation ring.

Now suppose that T is a valuation ring. Let $m = \cup m_{R_i} \hat{R}_i$ be the maximal ideal of T. Without loss of generality, we may assume that V/m_V is algebraic over k. For if this is not the case, we can replace k with a rational function field k' over k contained in all of the R_i such that V/m_V is algebraic over k'.

Suppose that for some index i, V_1 is a valuation ring with quotient field K_i which dominates $\hat{R}_i$ and V. We will show that there exists a valuation ring W_1 with quotient field K_∞ which dominates T and such that $W_1 \cap K_i = V_1$.

Consider the domain $A = T[V_1] \subset K_\infty$. Let $I \subset A$ be the ideal $I = (m + m_{V_1})A$. We will first establish that $1 \notin I$. If it were true that $1 \in I$, then there would exist an index $j > i$ such that $1 \in (m_{R_j} + m_{V_1})\hat{R}_j[V_1]$. Since $R_i \to R_j$ is birational, there exists an ideal $b \subset R_i$ and $x \in b$ such that $\frac{b}{x} \subset V$ and R_j is a localization of $R_i[\frac{b}{x}]$ at a maximal ideal. Thus $\hat{R}_i[\frac{b}{x}] \subset V_1$ since $V \subset V_1$. Let $\overline{\nu}$ be a valuation of K_i which has V_1 for its valuation ring. We have $\overline{\nu}(\hat{R}_i[\frac{b}{x}]) \geq 0$ and $\overline{\nu}(m_{R_j}) > 0$ since $m_{R_j} = m_{V_1} \cap R_j$. Thus $m_{R_j} \hat{R}_i[\frac{b}{x}] \subset m_{V_1}$, and V_1 dominates $\hat{R}_i[\frac{b}{x}]_{m_{R_j} \hat{R}_i[\frac{b}{x}]}$. Thus there exists a valuation ring U_1 of K_j such that $U_1 \cap K_i = V_1$ and U_1 dominates $\hat{R}_j$ (as follows from page 177 of [14]). $(m_{R_j} + m_{V_1})\hat{R}_j[V_1] \subset m_{U_1}$ implies $1 \notin (m_{R_j} + m_{V_1})\hat{R}_j[V_1]$. Thus we have a contradiction, and $1 \notin I$.

Let a be a prime ideal in A which contains I. Suppose that $h \in A/a$. There exists an index j such that we can write h as a class $h = [\sum f_r g_r]$ with $f_r \in V_1$ and $g_r \in \hat{R}_j$. We have natural inclusions $\hat{R}_j/m_{\hat{R}_j} \to A/a$ and $V_1/m_{V_1} \to A/a$ such that h is the image of the induced map $\hat{R}_j/m_{\hat{R}_j} \hat{\otimes}_k V_1/m_{V_1} \to A/a$. Thus h is algebraic over V_1/m_{V_1}, since $\hat{R}_j/m_{\hat{R}_j}$ is finite over k. We conclude that A/a is algebraic over V_1/m_{V_1}.

There exists a valuation ring W_1 which contains A such that $m_{W_1} \cap A = a$ and W_1/m_{W_1} is algebraic over A/a, by Corollary 3 to Theorem 5' of Section 4, Chapter VI [24]. Let $\overline{W}_1 = W_1 \cap K_i$. $\overline{W}_1$ contains V_1 and $m_{V_1} \subset m_{\overline{W}_1}$. But $\overline{W}_1/m_{\overline{W}_1}$ is algebraic over V_1/m_{V_1} since W_1/m_{W_1} is algebraic over V_1/m_{V_1}. Thus $V_1 = \overline{W}_1$ by Theorem 2 of Section 3, Chapter VI [24] and Corollary 1 to Theorem 5 of Section 4, Chapter VI [24].

We have thus proved the existence of an extension W_1 of V_1 to K_∞ which restricts to V_1 and dominates T.

Continuing with the proof of the theorem, suppose that for some index i, the extension of V to K_i which dominates $\hat{R}_i$ is not unique. There are then extensions V_1 and V_2 of V to K_i which dominate $\hat{R}_i$ such that $V_1 \not\subset V_2$ and $V_2 \not\subset V_1$.

We have shown that there then exist valuation rings W_1 and W_2 of K_∞ such that $T \subset W_1 \cap W_2$, $W_1 \cap K_i = V_1$ and $W_2 \cap K_i = V_2$. Thus $W_1 \not\subset W_2$ and $W_2 \not\subset W_1$. But this is impossible since T is a valuation ring of K_∞, by Theorem 3, Section 3, Chapter VI [24]. □

Corollary 4.5. *T is a valuation ring if for all i there does not exist an extension of V to K_i which dominates $\hat{R}_i$ of higher rank than the rank of V.*

Proof. This follows from Theorem 4.4 and the remark on page 181 of [14] which shows that if the extension of V to K_i which dominates $\hat{R}_i$ is not unique then there must be an extension of higher rank. □

The converse to the above corollary is false, as is seen by the following simple example. Let $p(t) \in k[[t]]$ be a transcendental power series with constant term zero. Consider the rank 1 discrete valuation ν on $k(x, y)$ defined by the embedding of k-algebras

$$k(x, y) \to k \ll t \gg$$

generated by $x = t, y = p(t)$ where $k \ll t \gg$ denotes the quotient field of $k[[t]]$. ν dominates $R = k[x, y]_{(x,y)}$. The valuation ring V of ν extends uniquely to a rank 2 valuation ring of the quotient field of $k[[x, y]]$ which dominates $k[[x, y]]$. Furthermore, the construction gives a unique extension of V to a rank 2 valuation ring which dominates $Q(\hat{S})$ for any algebraic normal local ring S of $k(x, y)$ such that V dominates S and S dominates R.

It follows from Theorem 4.4 that the examples in [14] of valuation rings dominating regular local rings R which do not have unique extensions in $Q(\hat{R})$ dominating $\hat{R}$ generate examples where T is not a valuation ring.

5. The prime ideal of elements of infinite value

We will assume in this section that V has rank 1, that is, the value group of V is a (possibly nondiscrete) subgroup Γ of $\mathbf{R}$. Other notations and assumptions will be as in Section 4.

Lemma 5.1. *Suppose that V has rank 1 and that R is an algebraic normal local ring of K such that V dominates R and $f \in \hat{R}$. Then one of the following must hold.*

1. *There exists $\rho \in \Gamma$ such that if $\{f_n\}$ is any Cauchy sequence in R which converges to f, then $\nu(f_n) = \rho$ for all $n \gg 0$.*
2. *If $\rho \in \Gamma$ and if $\{f_n\}$ is any Cauchy sequence in R which converges to f, then $\nu(f_n) > \rho$ for $n \gg 0$.*

Proof. We first argue that (1) or (2) must hold for a fixed Cauchy sequence $\{f_n\}$ in R which converges to f. Suppose that (2) doesn't hold. Then there exists $\overline{\rho} \in \Gamma$ such that given $n_0 \in \mathbf{N}$, there exists $\overline{n} > n_0$ such that $\nu(f_{\overline{n}}) \leq \overline{\rho}$. Let $i \in \mathbf{N}$ be such that $i\nu(m_R) > \overline{\rho}$, and let n_0 be such that $f_m - f_n \in m_R^i$ if $m, n > n_0$. There exists $\overline{n} > n_0$ such that $\nu(f_{\overline{n}}) \leq \overline{\rho}$. Then $\nu(f_n) = \nu(f_{\overline{n}})$ if $n > n_0$, so (1) holds for $\{f_n\}$.

If $\{f_n\}$ and $\{g_l\}$ are two distinct Cauchy sequences in R which converge to f, then for all $i \in \mathbf{N}$, there exists $n(i)$ such that $f_n - g_l \in m_R^i$ if $n, l \geq n(i)$. Thus (1)(or (2)) holds for $\{f_n\}$ if and only if (1) (or (2)) holds for $\{g_l\}$. □

Definition 5.2. *Let R be as in the statement of Lemma 5.1. Let*
$p(\hat{R})_\infty =$
$\{f\in\hat{R}\mid$ *(2) of Lemma 5.1 holds for a Cauchy sequence* $\{f_n\}$ *in R which converges to* $f\}$

Lemma 5.3. *Let R be as in the statement of Lemma 5.1. Then*

(1) *$p(\hat{R})_\infty$ is a prime ideal of $\hat{R}$ such that $p(\hat{R})_\infty\cap R=(0)$.*
(2) *There exists a unique extension $\overline{\nu}$ of ν to the quotient field $\overline{K}$ of $\hat{R}/p(\hat{R})_\infty$ which dominates $\hat{R}$. Let $\overline{V}$ be the valuation ring of $\overline{\nu}$. Then $(\overline{V},\overline{K})$ is an immediate extension of (V,K). That is, $\Gamma_{\overline{V}}=\Gamma_V$ and $k(V)=k(\overline{V})$.*

Proof. The facts that $p(\hat{R})_\infty$ is prime and $p(\hat{R})_\infty\cap R=(0)$ are immediate from Lemma 5.1. Suppose that $0\neq f+p(\hat{R})_\infty\in\hat{R}/p(\hat{R})_\infty$. We can find a Cauchy sequence $\{f_n\}$ in R such that $\{f_n\}$ satisfies (1) of Lemma 5.1 and $\{f_n\}$ converges to f. Let $\rho=\nu(f_n)$ for $n>>0$. We necessarily have that $\overline{\nu}(g)=\rho$ if $g=f+p(\hat{R})_\infty$. □

By a classical abuse of notation, we will say that $\nu(f)=\infty$ if $f\in p(\hat{R})_\infty$.

6. Simultaneous resolution of p_∞

Definition 6.1. *Suppose that R is a normal local ring which is essentially of finite type over a field k of characteristic zero, with quotient field K. A normal uniformizing transformation sequence (NUTS) is a sequence of ring homomorphisms*

$$\begin{array}{ccccccccccc}
R & \rightarrow & \overline{T}''_0 & \rightarrow & \overline{T}_0 & & & & & & \\
 & & \downarrow & & & \searrow & & & & & \\
 & & \overline{T}'_1 & \rightarrow & \overline{T}''_1 & \rightarrow & \overline{T}_1 & & & & \\
 & & & & \downarrow & & & \searrow & & & \\
 & & & & \overline{T}'_2 & \rightarrow & \overline{T}''_2 & \rightarrow & \overline{T}_2 & & \\
 & & & & & & \downarrow & & & \searrow & \\
 & & & & & & \vdots & & & \vdots & \\
 & & & & & & & & & & \downarrow \\
 & & & & & & & & & & \overline{T}'_n \;\rightarrow\; \overline{T}''_n \;\rightarrow\; \overline{T}_n
\end{array} \tag{2}$$

such that $\overline{T}_0=\hat{R}$, the completion of R with respect to its maximal ideal, and for all i, $\overline{T}_i$ is the completion with respect to its maximal ideal of a birational extension $\overline{T}'_i$ of $\overline{T}''_{i-1}$. For all i, $\overline{T}'_i$ is a normal local ring, $\overline{T}''_i$ is a normal local ring, essentially of finite type over $\overline{T}'_i$ with quotient field K_i such that $\overline{T}'_i\subset\overline{T}''_i\subset\overline{T}_i$ and K_0 is a finite extension of K, K_{i+1} is a finite extension of K_i for all $i\geq 0$.

Definition 6.1 is the extension of the definition of a UTS in Chapter 3 of [6] to normal local rings.

To simplify notation, we will often denote the NUTS (2) by $(R,\overline{T}''_n,\overline{T}_n)$ or by
$$R\rightarrow\overline{T}_0\rightarrow\overline{T}_1\rightarrow\overline{T}_n.$$
We will denote the NUTS consisting of the maps
$$\begin{array}{ccccccc}
\overline{T}'_{n-1} & \rightarrow & \overline{T}''_{n-1} & \rightarrow & \overline{T}_{n-1} & & \\
 & & \downarrow & & & \searrow & \\
 & & \overline{T}'_n & \rightarrow & \overline{T}''_n & \rightarrow & \overline{T}_n
\end{array}$$
by $\overline{T}_{n-1}\rightarrow\overline{T}_n$.

Suppose that ν is a rank 1 k-valuation of K, and R is dominated by ν. Suppose that $\overline{\nu}$ is an extension of ν to the quotient field of $\overline{T}_n$ which dominates $\overline{T}_n$. Then we will say that $\overline{T}_0 \to \overline{T}_n$ is a NUTS along ν. When there is no danger of confusion, we will denote $\overline{\nu}$ by ν.

We define

$$p(\overline{T}_n)_\infty = \{f \in \overline{T}_n \mid \overline{\nu}(f) = \infty\},$$
$$\lambda(\overline{T}_n'') = \lambda(\overline{T}_n) = \dim \overline{T}_n / p(\overline{T}_n)_\infty.$$

We define

$$\lambda(R) = \lambda(\overline{T}_0) = \dim \hat{R}/p(\hat{R})_\infty.$$

Lemma 6.2. *Suppose that*

$$\overline{T} \to \overline{T}(1) \to \cdots \to \overline{T}(t)$$

is a NUTS along ν. Then $\lambda(\overline{T}) \geq \lambda(\overline{T}(t))$.

Lemma 6.2 is the generalization of Lemma 6.3 of [8] to a NUTS. The proof is the same.

Let V be the valuation ring (in K) of ν and let

$$\lambda_V = \min \{\lambda(R) \mid R \text{ is a normal algebraic local ring of } K \text{ which is dominated by } V\}.$$

Theorem 6.3. *There exists an algebraic regular local ring $\overline{R}$ of K such that if $(R_1, \overline{T}''(t), \overline{T}(t))$ is a NUTS along ν with R_1 an algebraic normal local ring of K such that R_1 dominates $\overline{R}$, then*

$$\lambda(\overline{T}(t)) = \lambda_V.$$

Proof. Suppose that R is an algebraic normal local ring such that $\lambda(R) = \lambda_V$. Let $\overline{R}$ be an algebraic regular local ring of K such that $\overline{R}$ dominates R and V dominates $\overline{R}$. $\lambda(\overline{R}) = \lambda_V$ by Lemma 6.2. Suppose that $(R_1, \overline{T}''(t), \overline{T}(t))$ is a NUTS along ν with, R_1 an algebraic normal local ring of K such that R_1 dominates $\overline{R}$. We have $\lambda(R_1) = \lambda_V$ and $\lambda(\overline{T}(t)) \leq \lambda_V$ by Lemma 6.2. Let L be the quotient field of $\overline{T}''(t)$. L is a finite extension of K. By Theorem 4.2 [9] there exists an algebraic normal local ring R_3 of K and an algebraic regular local ring T_2 of L such that our extension of V to L dominates T_2, T_2 dominates $\overline{T}''(t)$, T_2 dominates R_3 and R_3 dominates R_1, with the property that $\hat{T}_2$ is finite over $\hat{R}_3$. Since $p(\hat{T}_2)_\infty \cap \hat{R}_3 = p(\hat{R}_3)_\infty$, we have that $\lambda(R_3) = \lambda(T_2)$. By Lemma 6.2,

$$\lambda_V = \lambda(R_1) \geq \lambda(\overline{T}''(t)) \geq \lambda(T_2) = \lambda(R_3) = \lambda_V.$$

□

Theorem 6.4. *Suppose that K^* is a finite field extension of K and ν^* is an extension of ν to K^*. Then $\lambda_{V^*} = \lambda_V$ and there exists an algebraic regular local ring $\tilde{R}$ of K such that the conclusions of Theorem 6.3 hold with $\overline{R} = \tilde{R}$ and if S is an algebraic normal local ring of K^* which is dominated by V^* and dominates $\tilde{R}$, and $(S, \overline{T}''(t), \overline{T}(t))$ is a NUTS along ν^*, then $\lambda(\overline{T}(t)) = \lambda_V$.*

Proof. Let $\overline{R}$ be the regular local ring of the conclusions of Theorem 6.3. Let S_1 be a normal algebraic local ring of K^* such that V^* dominates S_1 and $\lambda(S_1) = \lambda_{V^*}$. Let S_2 be an algebraic regular local ring such that V^* dominates S_2, S_2 dominates S_1 and S_2 dominates $\overline{R}$. By Lemma 6.2, $\lambda(S_2) = \lambda_{V^*}$. By Theorem 4.2 [9] there exists an algebraic normal local ring R_1 of K and an algebraic regular local ring S_3 of K^* such that V^* dominates S_3, S_3 dominates S_2, S_3 dominates R_1 and R_1 dominates $\overline{R}$, with the property that $\hat{S}_3$ is finite over $\hat{R}_1$. Since $p(\hat{S}_3)_\infty \cap \hat{R}_1 = p(\hat{R}_1)_\infty$, $\lambda(R_1) = \lambda(S_3)$. By Theorem 6.3 and Lemma 6.2,

$$\lambda_V = \lambda(R_1) = \lambda(S_3) = \lambda(S_2) = \lambda_{V^*}.$$

By Lemma 5.3 [9] there exists an algebraic regular local ring $\tilde{R}$ of K such that V dominates $\tilde{R}$, $\tilde{R}$ dominates $\overline{R}$ and if S is an algebraic normal local ring of K^* which is dominated by V^* and which contains $\tilde{R}$, then S dominates S_2.

By Theorem 6.3 applied to $S \subset K^*$ which dominates S_2, the conclusions of Theorem 6.4 hold. □

We now state a generalization of Theorem 5.1 [6] which resolves the prime ideal of infinite value terms.

Theorem 6.5. *Let k be a field of characteristic zero, K an algebraic function field, K^* a finite algebraic extension of K, ν^* a k-valuation of K^*, $\nu = \nu^* \mid K$, such that rank $\nu = 1$, rat rank $\nu = s$ and*

$$s \le \lambda = \lambda_V \le n,$$

where V is the valuation ring of ν and $n = \mathrm{trdeg}_k K - \mathrm{trdeg}_k V/m_V$. Suppose that S^* is an algebraic local ring with quotient field K^* which is dominated by ν^* and R^* is an algebraic local ring with quotient field K which is dominated by S^*. Let V^* be the valuation ring of ν^*. Then there exists a commutative diagram

$$\begin{array}{ccccc} R_0 & \rightarrow & S & \subset V^* \\ \uparrow & & \uparrow & \\ R^* & \rightarrow & S^* & \end{array}$$

where $S^* \rightarrow S$ and $R^* \rightarrow R_0$ are sequences of monoidal transforms along ν^* such that R_0 has regular parameters $(x_1, \ldots, x_n)$ and S has regular parameters $(y_1, \ldots, y_n)$ such that there are units $\delta_1, \ldots, \delta_s \in S$ and a $s \times s$ matrix $A = (a_{ij})$ of natural numbers such that $\det(A) \neq 0$,

$$\begin{aligned} x_1 &= y_1^{a_{11}} \cdots y_s^{a_{1s}} \delta_1 \\ &\vdots \\ x_s &= y_1^{a_{s1}} \cdots y_s^{a_{ss}} \delta_s \\ x_{s+1} &= y_{s+1} \\ &\vdots \\ x_n &= y_n \end{aligned} \tag{3}$$

and $\{\nu(x_1), \ldots, \nu(x_s)\}$, $\{\nu(y_1), \ldots, \nu(y_s)\}$ are rational bases of $\Gamma_\nu \otimes \mathbf{Q} = \Gamma_{\nu^*} \otimes \mathbf{Q}$. Furthermore,

$$p(\hat{R}_0)_\infty = (g_1, \ldots, g_{n-\lambda})$$

with

$$g_i \equiv x_{s+i} \bmod m(\hat{R}_0)^2$$

for $1 \le i \le n - \lambda$, and

$$p(\hat{S})_\infty = p(\hat{R}_0)_\infty \hat{S}$$

are regular primes.

Remark 6.6. *Suppose that in the hypothesis of Theorem 6.5 we further assume that $R^* \rightarrow S^*$ is such that R^* and S^* have regular parameters $(x_1^*, \ldots, x_n^*)$, $(y_1^*, \ldots, y_n^*)$ satisfying (3) and such that $\{\nu(x_1^*), \ldots, \nu(x_n^*)\}$, $\{\nu(y_1^*), \ldots, \nu(y_n^*)\}$ are rational bases of $\Gamma_\nu \otimes \mathbf{Q} = \Gamma_{\nu^*} \otimes \mathbf{Q}$. Then in the conclusions of Theorem 6.5 we further have that there exist $b_j(i) \in \mathbf{N}$, units $\alpha_i \in R_0$ such that*

$$x_i^* = x_1^{b_1(j)} \cdots x_s^{b_s(j)} \alpha_i$$

for $1 \le i \le s$ and there exist $c_j(i) \in \mathbf{N}$, units $\beta_i \in S$ such that

$$y_i^* = y_1^{c_1(j)} \cdots y_s^{c_s(j)} \beta_i$$

for $1 \le i \le s$.

This follows since all transformations in the proof of Theorem 6.5 are "CUTS in the first n *variables" (page* 49 [6]*).*

Proof. (of Theorem 6.5). Let $\tilde{R}$ be the regular local ring of the conclusions of Theorem 6.4. We first construct a commutative diagram

$$\begin{array}{ccc} R_1 & \rightarrow & S_1 \subset V^* \\ \uparrow & & \uparrow \\ R^* & \rightarrow & S^* \end{array}$$

such that the conclusions of Theorem 5.1 [6] hold, and R_1 dominates $\tilde{R}$. Let $R = R_1$, $\overline{T}'' = R_1$ and $\overline{T} = \hat{R}_1$. We will now show that we can construct a CUTS $\overline{T} \rightarrow \overline{T}(t)$ along ν, which is in the first n variables (with the notation of Theorem 4.7 of [6]), such that $p(\overline{T}(t))_\infty$ has the form of (53) of page 49 of [6],

$$\begin{aligned} p(\overline{T}(t))_\infty = (&\overline{z}_{r(1)}(t) - Q_{r(1)}(\overline{z}_1(t), \ldots, \overline{z}_{r(1)-1}(t)), \ldots, \\ &\overline{z}_{r(n-\lambda)}(t) - Q_{r(n-\lambda)}(\overline{z}_1(t), \ldots, \overline{z}_{r(n-\lambda)-1}(t))) \end{aligned} \tag{4}$$

with $s < r(1) < r(2) < \cdots < r(n-\lambda) \le n$ and such that for $1 \le i \le n - \lambda$,

$$Q_{r(i)} = \overline{z}_1(t)^{a_1(r(i))} \cdots \overline{z}_s(t)^{a_s(r(i))} u_{r(i)}$$

where $u_{r(i)}$ is a unit series in $\overline{z}_1(t), \ldots, \overline{z}_{r(i)}(t)$ with coefficients in $k(c_0, \ldots, c_t)$ (with the notation of (53) of page 49 of [6]).

The construction of $\overline{T} \rightarrow \overline{T}(t)$ follows from the proof of (53) of [6], with the insertion of the following at the bottom of page 54. "Since $\nu(Q_{\overline{m}}) = \nu(\overline{z}_{\overline{m}}(t)) < \infty$ we can perform by (54) [6] a UTS in the first $\overline{m} - 1$ variables to get $Q_{\overline{m}} = \overline{z}_1(t')^{a_1} \cdots \overline{z}_s(t')^{a_s} u_{\overline{m}}$ where $u_{\overline{m}} \in k(c_0, \ldots, c_{t'})[[\overline{z}_1(t'), \ldots, \overline{z}_{\overline{m}-1}(t')]]$ is a unit series".

Set $S = S_1$, $\overline{U}'' = S_1$, $\overline{U} = \hat{S}$. We can now construct a CUTS $\overline{U} \rightarrow \overline{U}(t')$ so that $(R_1, \overline{T}''(t'), \overline{T}(t'))$ and $(S_1, \overline{U}''(t'), \overline{U}(t'))$ is a CUTS along ν^*, by Lemma 4.3 and Lemma 4.4 [6].

Set $\tilde{g}_j(\overline{z}_1(t'), \ldots, \overline{z}_{r(j)}(t')) = \overline{z}_{r(j)}(t') - Q_{r(j)}(\overline{z}_1(t'), \ldots, \overline{z}_{r(j)-1}(t'))$ for $1 \le j \le n - \lambda$.

We will now show that the strict transform of $p(\hat{R}_1)_\infty$ in $\overline{T}(t')$ is $p(\overline{T}(t'))_\infty$. It suffices to show that the strict transform of $p(\hat{R}_1)_\infty$ in $\overline{T}(1)$ is $p(\overline{T}(1))_\infty$. Then the result follows by induction on t'. Let $p = p(\hat{R}_1)_\infty$. There exists an ideal I in R_1, $f \in I$, and a maximal ideal n in $\hat{R}_1[\frac{I}{f}]$ such that $\overline{T}(1) = \widehat{R_1[\frac{I}{f}]_n}$. Let

$$\overline{p} = \cup_{j=1}^{\infty} \left(p\hat{R}_1[\frac{I}{f}]_n : I^j \hat{R}_1[\frac{I}{f}]_n \right)$$

be the strict transform of p in $\hat{R}_1[\frac{I}{f}]_n$. $\overline{p} \neq \hat{R}_1[\frac{I}{f}]_n$ since the strict transform in $\hat{R}_1[\frac{I}{f}]_n$ of an element of infinite value must have infinite value. $\overline{p}$ is a prime ideal in $\hat{R}_1[\frac{I}{f}]_n$, and

$$\hat{R}_1/p \rightarrow \hat{R}_1[\frac{I}{f}]_n/\overline{p}$$

is birational (Section 0.2 [13], Corollary II.7.15 [12]). Thus $\dim \overline{T}(1)/\overline{p}\overline{T}(1) = \lambda$. $\overline{p}\overline{T}(1)$ is a prime contained in $p(\overline{T}(1))_\infty$ and $\dim \overline{T}(1)/p(\overline{T}(1))_\infty = \lambda$ by Theorem 6.4 (since R_1 contains $\tilde{R}$). Thus $\overline{p}\overline{T}(1) = p(\overline{T}(1))_\infty$.

Hence there exist $f_1, \ldots, f_{n-\lambda} \in p(\hat{R}_1)_\infty$, $c_1(i), \ldots, c_s(i) \in \mathbf{N}$ for $1 \le i \le n - \lambda$ and $b_{ij} \in \overline{T}(t')$ such that $\det(b_{ij})$ is a unit in $\overline{T}(t')$, $f_i = M_i(\sum_{j=1}^{n-\lambda} b_{ij}\tilde{g}_j)$ where

$M_i = \overline{z}_1(t')^{c_1(i)} \cdots \overline{z}_s(t')^{c_s(i)}$ for $1 \le i \le n-\lambda$ and

$$p(\overline{T}(t'))_\infty = \left(\frac{f_1}{M_1}, \dots, \frac{f_{n-\lambda}}{M_{n-\lambda}}\right).$$

Let m be a positive integer such that

$$m > \frac{(\max_{1\le j\le n-\lambda}\nu(\overline{z}_{r(j)}(t'))) + (\max_{1\le i\le n-\lambda}\nu(M_i))}{\nu(m(\overline{T}(t')))}. \tag{5}$$

By Theorem 4.8 [6] (with $l = n$) there exists a CRUTS along ν, $(R_1, R_1, T(t'))$ and $(S_1, S_1, U(t'))$ with associated MTSs

$$\begin{array}{ccc} S & \to & S(t') \\ \uparrow & & \uparrow \\ R & \to & R(t') \end{array}$$

such that (with the notation of Theorem 4.8 [6])

$$f_i = \overline{M}_i(\sum_{j=1}^{n-\lambda} b_{ij}(\tilde{\overline{x}}_1(t'), \dots, \tilde{\overline{x}}_n(t'))\tilde{g}_j(\tilde{\overline{x}}_1(t'), \dots, \tilde{\overline{x}}_{r(j)}(t')) + h_i \in p(\hat{R}(t'))_\infty$$

with $\overline{M}_i = \tilde{\overline{x}}_1(t')^{c_1(i)} \cdots \tilde{\overline{x}}(t')^{c_s(i)}$, $h_i \in m(T(t'))^m$ (where m is the integer of (5)) for $1 \le i \le n-\lambda$, and such that (by (A3) of page 83 of [6]) $\nu(\overline{z}_i(t')) = \nu(\tilde{\overline{x}}_i(t'))$ for $1 \le i \le n$, and

$$\begin{aligned} \nu(\overline{z}_{r(i)}(t')) &= \nu(Q_{r(i)}(\overline{z}_1(t'), \dots, \overline{z}_{r(i)-1}(t'))) \\ &= \nu(Q_{r(i)}(\tilde{\overline{x}}_1(t'), \dots, \tilde{\overline{x}}_{r(i)-1}(t'))) \\ &= \nu(\tilde{\overline{x}}_{r(i)}(t')) \end{aligned}$$

for $1 \le i \le n-\lambda$.

Now we perform the MTS

$$\begin{array}{ccc} S(t') & \to & S(t'') \\ \uparrow & & \uparrow \\ R(t') & \to & R(t'') \end{array}$$

of the proof of Theorem 4.9 [6] (with $l = n$). Because of the form of the $\tilde{g}_j$, we have for $1 \le j \le n-\lambda$,

$$\tilde{g}_j = \overline{x}_1(t'')^{b_1(j)} \cdots \overline{x}_s(t'')^{b_s(j)}\overline{g}_j$$

where

$$\overline{g}_j \equiv \tilde{d}_j \overline{x}_{r(j)}(t'') \bmod (\overline{x}_1(t''), \dots, \overline{x}_{r(j)-1}(t''))\hat{R}(t'') + m(\hat{R}(t''))^2$$

for some nonzero $\tilde{d}_j \in R(t'')/m(R(t''))$, where

$$\nu(\overline{x}_1(t'')^{b_1(j)} \cdots \overline{x}_s(t'')^{b_s(j)}) = \nu(\tilde{\overline{x}}_{r(j)}(t')) = \nu(\overline{z}_{r(j)}(t')). \tag{6}$$

We further have

$$\overline{h}_i = \frac{h_i}{\overline{M}_i} = \overline{x}_1(t'')^{e_1(i)} \cdots \overline{x}_s(t'')^{e_s(i)}\tilde{h}_i$$

for $1 \le i \le n-\lambda$ with $\tilde{h}_i \in \hat{R}(t')$ and

$$\nu(\overline{x}_1(t'')^{e_1(i)} \cdots \overline{x}_s(t'')^{e_s(i)}) > \max\ \nu(\overline{z}_{r(j)}(t)) \tag{7}$$

for $1 \le j \le n-\lambda$ by (5). By Lemma 4.2 [6], (6) and (7), we can further choose the final CUTS of type (M1) (on the top of page 89 of [6]) so that

$$e_j(i) > \max_{1\le\alpha\le n-\lambda} b_j(\alpha)$$

for $1 \le j \le s$ and $1 \le i \le n-\lambda$.

Let $A = (b_{ij})^{-1}$, a matrix with coefficients in $\hat{R}(t'')$.

$$A\begin{pmatrix} \frac{f_1}{M_1} \\ \vdots \\ \frac{f_{n-\lambda}}{M_{n-\lambda}} \end{pmatrix} = \begin{pmatrix} \tilde{g}_1 + d_1 \\ \vdots \\ \tilde{g}_{n-\lambda} + d_{n-\lambda} \end{pmatrix}$$

where for $1 \le i \le n-\lambda$,

$$d_i = \overline{x}_1(t'')^{b_1(i)+1} \cdots \overline{x}_s(t'')^{b_s(i)+1}\overline{d}_i$$

for some $\overline{d}_i \in \hat{R}(t'')$. Thus

$$g_i = \overline{g}_i + \overline{d}_i \in p(\hat{R}(t''))_\infty$$

and

$$g_i \equiv \tilde{d}_i \overline{x}_{r(i)}(t'') \bmod (\overline{x}_1(t''), \ldots, \overline{x}_{r(i)-1}(t''))\hat{R}(t'') + m(\hat{R}(t''))^2.$$

Thus $(g_1, \ldots, g_{n-\lambda})$ is a complete intersection and a regular prime ideal in $\hat{R}(t'')$. Since $\lambda(\hat{R}(t'')) = \lambda$ (by Theorem 6.3), we have that $(g_1, \ldots, g_{n-\lambda})$ is a basis of $p(\hat{R}(t''))_\infty$. Since $(g_1, \ldots, g_{n-\lambda})\hat{S}(t'')$ is a prime ideal and $\lambda(S(t'')) = \lambda$ (by Theorem 6.4), it follows that $p(\hat{R}(t''))_\infty \hat{S}(t'') = p(\hat{S}(t''))_\infty$.

We can now make a change of variables in the regular parameters $(x_1(t''), \ldots, x_n(t''))$ and $(y_1(t''), \ldots, y_n(t''))$ of the proof of Theorem 4.9 [6] to get the desired forms of the g_i.

$\{\nu(x_1), \ldots, \nu(x_s)\}$ and $\{\nu^*(y_1), \ldots, \nu^*(y_s)\}$ are rational bases of $\Gamma^* \otimes \mathbf{Q}$ by the construction of the sequence $R^* \to R_0$ and $S^* \to S$. □

7. Rank 1 valuations

Let notations be as in Section 4. Further assume that V has rank 1. Consider our directed set $\{R_i \mid i \in I\}$ satisfying (a), (b), (c) and (d). For $i \in I$, we define

$$p_i = p(\hat{R}_i)_\infty = \{f \in \hat{R}_i \mid \nu(f) = \infty\}.$$

For $i < j$, the natural inclusions $\hat{R}_i \to \hat{R}_j$ induce inclusions $\hat{R}_i/p_i \to \hat{R}_j/p_j$. Thus $\{\hat{R}_i/p_i \mid i \in I\}$ is a directed system, and we have a local domain

$$\overline{T} = \lim_{\longrightarrow} \hat{R}_i/p_i = \cup \hat{R}_i/p_i.$$

Let $\overline{K}_\infty$ be the quotient field of $\overline{T}$.

Lemma 7.1. *Suppose that V has rank 1. Then the ring $\overline{T} = \cup_{i\in I}\hat{R}_i/p_i$ does not depend on the directed system of rings $\{R_i \mid i \in I\}$ satisfying (a), (b), (c) and (d).*

Proof. The proof is essentially the same as that of Lemma 4.1. We must observe that the inclusion $\hat{R}_i \to \hat{S}_j$ of the proof of Lemma 4.1 induces a natural inclusion $\hat{R}_i/p_i \to \hat{S}_j/q_j$, where $q_j = \{f \in \hat{S}_j \mid \nu(f) = \infty\}$. □

Theorem 7.2 is a generalization of Zariski's local uniformization theorem [23]. Our proof is an extension in rank 1 of the proof for general rank in [9, 6.2]. We incorporate the conclusions of Theorem 6.5 which resolves the prime ideal of infinite value terms.

Theorem 7.2. *Let k be a field of characteristic zero, K an algebraic function field over k, and let ν be a rank 1 k-valuation of K, of rational rank s, with valuation ring V. Let*

$$n = trdeg_k K - trdeg_k V/m_V,$$

$\lambda = \lambda_V$ (defined before Theorem 6.3).

Then there exists a partially ordered set I and algebraic regular local rings $\{R_i \mid i \in I\}$ with quotient field K which are dominated by V such that

$$V = \lim_{\rightarrow} R_j = \cup_{j \in I} R_j$$

and R_j has regular parameters $(x_1(j), \ldots, x_n(j))$ such that

(1)
$$\{\nu(x_1(j)), \ldots, \nu(x_s(j)\}$$
is a rational basis of $\Gamma \otimes \mathbf{Q}$.

(2) *If $j < k \in I$ then there are relations*
$$x_i(j) = \prod_{l=1}^{s} x_l(k)^{d_{il}} \delta_i \tag{8}$$
for $1 \leq i \leq s$ where $\delta_i \in R_k$ are units. The $s \times s$ matrix $D(j,k) = (d_{il})$ of (8) has nonzero determinant.

(3) *The prime ideal*
$$p_j = p(\hat{R}_j)_\infty = \{f \in \hat{R}_j \mid \nu(f) = \infty\} = (g_1(j), \ldots, g_{n-\lambda}(j))$$
with
$$g_i(j) \equiv x_{s+i}(j) \ mod\ m(\hat{R}_j)^2.$$
In particular, p_j is a regular prime.

(4) *For $j \in I$, let Λ_j be the free $\mathbf{Z}$-module $\Lambda_j = \sum_{i=1}^{s} \nu(x_i(j))\mathbf{Z}$. Then*
$$\Gamma = \lim_{\rightarrow} \Lambda_j = \cup_{j \in I} \Lambda_j.$$

Proof. Let R^* be an algebraic regular local ring such that V dominates R^*. By Theorem 6.5 (with $K = K^*$ and $R^* = S^*$), there exists a sequence of monoidal transforms $R^* \to R_0$ along V such that (1) and (3) of this theorem hold on R_0.

Suppose that m is a positive integer and $f = (f_1, \ldots, f_m) \in V^m$. We will construct a sequence of monoidal transforms $R_0 \to R_f$ along V such that $f_1, \ldots, f_m \in R_f$, (1) and (3) of this theorem hold for R_f and (2) of this theorem holds for $R_0 \to R_f$. We will further have $\nu(f_1), \ldots, \nu(f_m) \in \Lambda_f$.

By Theorem 4.9 [9] with the R^*, S^* of the statement of Theorem 4.9 set as $R^* = S^* = R_0$, and $v_i = x_i(0)$ if $1 \leq i \leq s$, and $v_{s+1} = f_1, \ldots, v_{s+m} = f_m$, there exists a sequence of monoidal transforms $R_0 \to R_1$ along V such that (1) of this theorem holds for R_1, (2) of this theorem holds for $R_0 \to R_1$, $f_1, \ldots, f_m \in R_1$ and $\nu(x_1), \ldots, \nu(x_s) \in \Lambda_1$. By Theorem 6.5 (with $K = K^*$, $R^* = S^*$) and Remark 6.6, there exists a sequence of monoidal transforms $R_1 \to R_f$ along V such that (1), (2) and (3) of this theorem hold for $R_0 \to R_f$ and $\nu(f_1), \ldots, \nu(f_m) \in \Lambda_f$. We have $\det D(0,f) \neq 0$ since $\{\nu(x_i(0)) \mid 1 \leq i \leq s\}$ and $\{\nu(x_i(f)) \mid 1 \leq i \leq s\}$ are two bases of $\Gamma \otimes \mathbf{Q}$.

Let $I = \sqcup_{m \in \mathbf{N}_+} V^m$ be the disjoint union. For $f \in I$ we construct R_f as above. If $f = 0$ we let R_0 be the R_0 constructed above. Define a partial order on I by $f \leq g$ if $R_f \subset R_g$.

Suppose that $R_\alpha \subset R_\beta$. We have $R_0 \subset R_\alpha \subset R_\beta$.

$$x_i(0) = \prod_{j=1}^{s} x_j(\alpha)^{c_{ij}} \delta_i$$

for $1 \leq i \leq s$ with δ_i a unit in R_α and

$$x_i(0) = \prod_{j=1}^{s} x_j(\beta)^{d_{ij}} \epsilon_i$$

for $1 \le i \le s$ with ϵ_i a unit in R_β. Thus in R_β there are factorizations

$$x_i(\alpha) = \prod_{j=1}^{s} x_j(\beta)^{e_{ij}} \lambda_i$$

for $1 \le i \le s$ and λ_i a unit in R_β. We have $\det(D(\alpha,\beta)) \ne 0$ since (1) holds for R_α and R_β. Thus (2) holds for $R_\alpha \to R_\beta$. To show that $V = \lim_{\to} R_j$, we must verify that I is a directed set. That is, for $\alpha, \beta \in I$, there exists $\gamma \in I$ such that $R_\alpha \subset R_\gamma$ and $R_\beta \subset R_\gamma$.

There exists $f_1, \ldots, f_t \in V$ such that if $A = k[f_1, \ldots, f_t]$, $m = A \cap m_V$ then $R_\alpha = A_m$. There exists $g_1, \ldots, g_n \in V$ such that if $B = k[g_1, \ldots, g_n]$, $n = B \cap m_V$ then $R_\beta = B_n$. Set $\gamma = (f_1, \ldots, f_t, g_1, \ldots, g_n)$. By construction, $A, B \subset R_\gamma$. Since $m_V \cap R_\gamma = m_\gamma$ is the maximal ideal of R_γ, we have R_α, $R_\beta \subset R_\gamma$.

(4) holds by our construction, since $\nu(f) \in \Lambda_f$ if $f \in V$. □

Theorem 7.3. *Suppose that V has rank 1 and k has characteristic zero. Then the ring $\overline{T} = \cup_{i \in I}(\hat{R}_i/p_i)$ is a valuation ring.*

Proof. Let s denote the rational rank of ν. By Lemma 7.1 we can assume that the rings R_i's are as in Theorem 7.2. R_i has regular parameters $x_1(i), \ldots, x_n(i)$ and $\hat{R}_i = R_i/m_{R_i}[[x_1(i), \ldots, x_n(i)]]$.

Let $f \in \hat{R}_i$. We recall that if $\nu(f) < \infty$, then by Theorem 4.8 and Theorem 4.10 [6] after a MTS $R_i \to \overline{R}(1)$ along ν

$$f = \overline{x}_1(1)^{d_1} \cdots \overline{x}_s(1)^{d_s} u(\overline{x}_1(1), \ldots, \overline{x}_n(1))$$

where $\overline{x}(1), \ldots, \overline{x}_n(1)$ are regular parameters in the ring $\overline{R}(1)$, $\nu(\overline{x}(1)), \ldots, \nu(\overline{x}_s(1))$ are rationally independent and $u \in \widehat{\overline{R}(1)}$ is a unit power series. Further, there exist units $\alpha_j \in \overline{R}(1)$ such that

$$x_j(i) = \overline{x}_1(1)^{\beta_1(j)} \cdots \overline{x}_s(1)^{\beta_s(j)} \alpha_j$$

for $1 \le j \le s$.

Let $h \in Q(\overline{T})$. We want to show that either $h \in \overline{T}$, or $1/h \in \overline{T}$. So it suffices to show that if $\nu(h) \ge 0$, then $h \in \overline{T}$. Write $h = a/b$ where $a \in \cup_{i \in I}\hat{R}_i/p_i$ and $0 \ne b \in \cup_{i \in I}\hat{R}_i/p_i$. Then $a \in \hat{R}_j/p_j$ for some $j \in I$, and $b \in \hat{R}_k/p_k$ for some $k \in I$. After a MTS $R_j \to \overline{R}(1)$ along ν we have

$$a = \overline{x}_1(1)^{d_1} \cdots \overline{x}_s(1)^{d_s} u(\overline{x}_1(1), \ldots, \overline{x}_n(1))$$

where $\overline{x}_1(1), \ldots, \overline{x}_n(1)$ are regular parameters in the ring $\overline{R}(1)$ and $u \in \widehat{\overline{R}(1)}$ is a unit. Further, there exist units $\alpha_j \in \overline{R}(1)$ such that

$$x_j(i) = \overline{x}_1(1)^{\beta_1(j)} \cdots \overline{x}_s(1)^{\beta_s(j)} \alpha_j \tag{9}$$

for $1 \le j \le s$.

After another MTS $R_k \to \overline{R}(2)$ along ν we have

$$b = \overline{x}_1(2)^{e_1} \cdots \overline{x}_s(2)^{e_s} u'(\overline{x}_1(2), \ldots, \overline{x}_n(2))$$

where $\overline{x}_1(2), \ldots, \overline{x}_n(2)$ are regular parameters in the ring $\overline{R}(2)$ and $u' \in \widehat{\overline{R}(2)}$ is a unit. Further, there exist units $\gamma_j \in \overline{R}(2)$ such that

$$x_j(k) = \overline{x}_1(2)^{\delta_1(j)} \cdots \overline{x}_s(2)^{\delta_s(j)} \gamma_j \tag{10}$$

for $1 \le j \le s$.

We have that $\overline{R}(1) = k[f_1, \ldots, f_m]_{m_V \cap k[f_1,\ldots,f_m]}$ for some $f_1, \ldots, f_m \in V$, and $\overline{R}(2) = k[g_1, \ldots, g_n]_{m_V \cap k[g_1,\ldots,g_n]}$ for some $g_1, \ldots, g_n \in V$.

Let $c = (f_1, \ldots, f_m, g_1, \ldots, g_n) \in V^{m+n}$. Let R_c be constructed as in the proof of Theorem 7.2. Then $R_c \in \cup_{i \in I} R_i$ and $\overline{R}(1) \subset R_c$, $\overline{R}(2) \subset R_c$. The ring R_c has regular parameters $x_1(c), \ldots, x_n(c)$ and by (2) of Theorem 7.2 and (9) and (10) the "good form" of a and b is preserved in R_c:

$$a = x_1(c)^{a_1} \ldots x_s(c)^{a_s} \bar{u}(x_1(c), \ldots, x_n(c))$$

$$b = x_1(c)^{b_1} \ldots x_s(c)^{b_s} \bar{u}'(x_1(c), \ldots, x_n(c))$$

where $\bar{u}$ and $\bar{u}'$ are units in $\hat{R}_c$.

Let $g = x_1(c)^{a_1} \ldots x_s(c)^{a_s} / x_1(c)^{b_1} \ldots x_s(c)^{b_s}$. Since $g \in K = Q(R_c)$ and $\nu(g) = \nu(h) \geq 0$, we have that $g \in V$ and $g \in R_g$, which is in the directed system I. We have $g = x_1(g)^{f_1} \ldots x_s(g)^{f_s} w(x_1(g), \ldots x_n(g))$ with $f_i \geq 0$ for every i and w a unit in R_g. There exists $\gamma \in I$ such that $R_c \subset R_\gamma$ and $R_g \subset R_\gamma$. Then

$$h = x_1(\gamma)^{t_1} \ldots x_s(\gamma)^{t_s} \bar{w}(x_1(\gamma), \ldots, x_n(\gamma)),$$

with $t_i \geq 0$ for every i and $\bar{w}$ a unit in $\widehat{R}_\gamma$. Hence $h \in \widehat{R}_\gamma$ and so $h \in \overline{T}$. □

Theorem 7.4. *Suppose that V has rank 1 and k has characteristic zero. Then*

(1) *$(\overline{T}, Q(\overline{T}))$ is an Henselian immediate extension of (V, K).*

(2) *There exists a directed system of regular local rings $\{R_i\}$ satisfying (a), (b), (c), and (d) such that each $\hat{R}_i/p_i$ is a regular local ring, and $\overline{T} = \cup \hat{R}_i/p_i$.*

Proof. Let $\overline{K}_i$ be the quotient field of $\hat{R}_i/p_i$. Then for all $i \in I$, $(\overline{T} \cap \overline{K}_i, \overline{K}_i)$ is an immediate extension of (V, K) by Lemma 5.3. Thus $(\overline{T}, Q(\overline{T}))$ is an immediate extension of (V, K).

By an extension of Proposition 4.2, $\overline{T}$ is Henselian.

Statement (2) follows from the construction of Theorem 7.2. □

Suppose that W is a rank 1 valuation ring, with valuation ω. Let $\phi(x) = e^{-\omega(x)}$ for $x \in L = Q(W)$. A sequence $(x_i)_{i \in \mathbf{N}}$ of elements of L is ϕ-Cauchy if given $\epsilon > 0$, there exists n_0 such that $\phi(x_n - x_m) < \epsilon$ for all $m, n \geq n_0$ (Section 2, [10]). W is said to be complete if all ϕ-Cauchy sequences (x_i) converge to an $x \in L$.

Example 7.5. *Even if V has rank 1 and $T = \overline{T}$ is a valuation ring, (which necessarily has residue field $k(T) = k(V)$ and value group $\Gamma_V = \Gamma_T$), (T, K_∞) is not in general complete. In particular, it is not a maximal immediate extension (in the sense of Krull [18] and Kaplansky [16]).*

Proof. Let $K = k(x, y)$ be a rational function field in two variables over a field k of characteristic zero. Let $R = k[x, y]_{(x,y)}$. Let ν be the rank one valuation of K with nondiscrete value group which we can take to be $\mathbf{Q}$ and residue field k which dominates R constructed in Example 3, page 102 of [24]. Let

$$R \to R_1 \to R_2 \to \cdots \to R_i \to \cdots$$

be the system of regular local rings for $i \in \mathbf{N}$ of the construction such that $\cup R_i = V$ is the valuation ring of ν.

We will first establish that $T = \cup \hat{R}_i$ is a valuation ring with residue field $k(V)$ and value group $\mathbf{Q}$. For any fixed i, let $q_i = \{f \in R_i \mid \nu(f) = \infty\}$. By Lemma 5.3 ν extends uniquely to a valuation of $Q(\hat{R}_i/q_i)$ which dominates $\hat{R}_i/q_i$ and has residue field k and value group $\mathbf{Q}$. If $q_i \neq (0)$, then $\hat{R}_i/q_i$ is a 1 dimensional excellent local ring, so the only valuation rings of $Q(\hat{R}_i/q_i)$ which dominate $\hat{R}_i/q_i$ are discrete, which is a contradiction. Thus $q_i = 0$. By Theorems 7.3 and 7.4, $T = \overline{T}$ is a valuation ring with value group $\mathbf{Q}$ and residue field k.

In each regular local ring $\hat{R}_i$ there is the sequence of all valuation ideals

$$\cdots \subset I_n(i) \subset \cdots \subset I_1(i) \subset I_0(i) = m(\hat{R}_i) \subset \hat{R}_i.$$

Let $p_j(i) = \nu(I_j(i))$. For fixed j, $lim_{j\to\infty} p_j(i) = \infty$ (c.f. Lemma 2.3 [6]). Notice that $\cup_{i\in\mathbf{N}} \cup_{j\in\mathbf{N}} \{p_j(i)\} = \mathbf{Q}_+$, but arbitrarily large elements of $\mathbf{Q}_+$ are not in $\cup_{j\in\mathbf{N}}\{p_j(i)\}$ for a fixed i since $\cup_{j\in\mathbf{N}}\{p_j(i)\}$ is not discrete.

We can inductively construct for all $i \in \mathbf{N}$, $\sigma(i), \lambda(i) \in \mathbf{N}$ and $a_i \in I_{\lambda(i)}(\sigma(i))$ such that $i < \sigma(i)$, $\sigma(i-1) < \sigma(i)$, $\nu(a_i) = p_{\lambda(i)}(\sigma(i))$,

$$\nu(a_i) > \ \max\ \{p_{\lambda(0)}(\sigma(0)), \ldots, p_{\lambda(i-1)}(\sigma(i-1)), i\},$$

and $p_{\lambda(i)}(\sigma(i)) \notin \cup_{k<\sigma(i)} \cup_{j\in\mathbf{N}} \{p_j(k)\}$ for every i.

For $i \in \mathbf{N}$, set $\alpha_i = a_1 + \cdots + a_i$. For $i < j$, we have

$$\nu(\alpha_j - \alpha_i) = \nu(a_{i+1}) > i.$$

Thus $\{\alpha_i\}$ is a ϕ-Cauchy sequence. Suppose that there exists a limit $\tau \in K_\infty$ of $\{\alpha_i\}$. Then

$$\nu(\tau - \alpha_i) = p_{\lambda(i+1)}(\sigma(i+1)) \tag{11}$$

for all i, by the definition of a limit. We have $\tau \in T$ so that $\tau \in \hat{R}_{\sigma(i)}$ for some i. Thus $\tau - \alpha_i \in \hat{R}_{\sigma(i)}$. But by (11) we have that $\nu(\tau - \alpha_i)$ is not the value of an element of $\hat{R}_{\sigma(i)}$, a contradiction. □

8. Ramification of completions of rank 1 valuation rings

Theorem 8.1 is a generalization of Theorem 6.3 [9], which resolves the prime ideal of infinite value terms.

Theorem 8.1. *Let k be a field of characteristic zero, K an algebraic function field over k, K^* a finite algebraic extension of K, V^* a rank 1 k-valuation ring of K^* of rational rank s, $V = V^* \cap K$. Let*

$$e = [\Gamma^* : \Gamma]$$

be the ramification index of V^* relative to V,

$$f = [V^*/m_{V^*} : V/m_V]$$

be the residue degree of V^* relative to V, and let τ be a primitive element of V^*/m_{V^*} over V/m_V. Let

$$n = trdeg_k K^* - trdeg_k V^*/m_{V^*} = trdeg_k K - trdeg_k V/m_V,$$

$\lambda = \lambda_V = \lambda_{V^*}$ (as shown in Theorem 6.4).

Then there exists a partially ordered set I and algebraic regular local rings $\{S_i \mid i \in I\}$ with quotient field K^* which are dominated by V^* where S_j has regular parameters $(y_1(j), \ldots, y_n(j))$ such that

(1)
$$\{\nu^*(y_1(j)), \ldots, \nu^*(y_s(j))\}$$
is a rational basis of $\Gamma^* \otimes \mathbf{Q}$.

(2) *For all $k \in I$ there exist algebraic regular local rings $R_0(k)$ with quotient field K which are dominated by V such that there exist factorizations*

$$R_0(k) \to R_k \to S_k$$

so that there are regular parameters $(x_1(k), ..., x_n(k))$ in $R_0(k)$, units $\delta_1(k), \ldots, \delta_s(k) \in S_k$ and a $s \times s$ matrix $A(k) = (a_{ij}(k))$ of nonnegative integers such that $det(A(k)) \neq 0$ and

$$\begin{aligned} x_1(k) &= y_1(k)^{a_{11}(k)} \cdots y_s(k)^{a_{1s}(k)} \delta_1(k) \\ &\vdots \\ x_s(k) &= y_1(k)^{a_{s1}(k)} \cdots y_s(k)^{a_{ss}(k)} \delta_s(k) \\ x_{s+1}(k) &= y_{s+1}(k) \\ &\vdots \\ x_n(k) &= y_n(k). \end{aligned} \tag{12}$$

R_k is a normal local ring with quotient field K (which is obtained by a toric blowup of $R_0(k)$) such that S_k is a localization at a maximal ideal of the integral closure of R_k in K^*. The prime ideals

$$p_j = p(\hat{R}_j)_\infty = \{f \in \hat{R}_j \mid \nu(f) = \infty\} = (g_1(j), \ldots, g_{n-\lambda}(j))$$

with

$$g_i(j) \equiv x_{s+i}(j) \ mod \ m(\hat{R}_j)^2$$

and

$$q_j = p(\hat{S}_j)_\infty = p(\hat{R}_j)_\infty \hat{S}_j = (g_1(j), \ldots, g_{n-\lambda}(j))$$

with

$$g_i(j) \equiv y_{s+i}(j) \ mod \ m(\hat{S}_j)^2.$$

Furthermore, there are isomorphisms of abelian groups

$$\Gamma^*/\Gamma \cong \mathbf{Z}^s/A(k)\mathbf{Z}^s,$$

$$[S_k/m_{S_k} : R_k/m_{R_k}] = f, \ \mid det(A(k)) \mid = e, \ [QF(\hat{S}_k) : QF(\hat{R}_k)] = ef$$

and $S_k/m_{S_k} = R_k/m_{R_k}[\tau]$.

(3) *Let k' be an algebraic closure of V^*/m_{V^*}. Suppose that $j < k \in I$.*

(a) *There are relations*

$$y_i(j) = \prod_{c=1}^{s} y_c(k)^{d_{ic}} \epsilon_i \tag{13}$$

where d_{ic} are natural numbers and $\epsilon_i \in S_k$ is a unit for $1 \leq i \leq s$. Let $D(j,k)$ be the $s \times s$ matrix of (13). Then $det(D(j,k)) \neq 0$.

(b) *There exists a commutative diagram*

$$\begin{array}{ccc} R_k & \to & S_k \\ \uparrow & & \uparrow \\ R_j & \to & S_j \end{array} \tag{14}$$

(c) *We have actions of Γ^*/Γ on $\hat{S}_j \hat{\otimes}_{S_j/m_{S_j}} k'$ such that*

$$(\hat{S}_j \hat{\otimes}_{S_j/m_{S_j}} k')^{\Gamma^*/\Gamma} \cong \hat{R}_j \hat{\otimes}_{R_j/m_{R_j}} k'$$

for all j, and this action is compatible with restriction.

We have an isomorphism

$$\hat{S}_j \hat{\otimes}_{S_j/m_{S_j}} k' \cong k'[[\overline{y}_1(j), \ldots, \overline{y}_n(j)]]$$

where $\overline{y}_1(j), \ldots, \overline{y}_n(j)$ *are defined by*

$$x_i(j) = \begin{cases} \overline{y}_1(j)^{a_{i1}(j)} \cdots \overline{y}_s(j)^{a_{is}(j)} & \text{if } 1 \leq i \leq s \\ \overline{y}_i(j) & \text{if } s+1 \leq i \leq n \end{cases}$$

Let $(b_{\alpha,\beta}(j)) = adjA(j)$ *and* ω *be a primitive e-th root of unity. The action of* Γ^*/Γ *on* $\hat{S}_j \hat{\otimes}_{S_j/m_{S_j}} k' \cong k'[[\overline{y}_1(j), \ldots, \overline{y}_n(j)]]$ *is defined for*

$$c \in \mathbf{Z}^s / A(j)\mathbf{Z}^s \cong \Gamma^*/\Gamma$$

by

$$\sigma_c(\overline{y}_\alpha(j)) = \begin{cases} \omega^{\sum_{\beta=1}^s b_{\alpha,\beta}(j)c_\beta} \overline{y}_\alpha(j) & \text{if } 1 \leq \alpha \leq s \\ \overline{y}_\alpha(j) & \text{if } s+1 \leq j \leq n. \end{cases}$$

(4)

$$V^* = \lim_{\rightarrow} S_j = \cup_{j \in I} S_j$$

and

$$V = \lim_{\rightarrow} R_j = \cup_{j \in I} R_j.$$

For $j \in I$, *let* Λ_j *be the free* $\mathbf{Z}$ *module* $\Lambda_j = \sum_{i=1}^s \nu(x_i(j))\mathbf{Z}$, *and let* Ω_j *be the free* $\mathbf{Z}$ *module* $\Omega_j = \sum_{i=1}^s \nu^*(y_i(j))\mathbf{Z}$. *Then*

$$\Gamma = \lim_{\rightarrow} \Lambda_j = \cup_{j \in I} \Lambda_j$$

and

$$\Gamma^* = \lim_{\rightarrow} \Omega_j = \cup_{j \in I} \Omega_j.$$

Proof. Suppose that R' is the regular local ring of Theorem 6.1 [9], and $\tilde{R}$ is the regular local ring of Theorem 6.4. By Theorem 6.1 [9], there exists a sequence of local rings

$$R_0(0) \rightarrow R_0 \rightarrow S_0$$

such that $R' \subset R_0(0)$, $\tilde{R} \subset R_0(0)$ and the conclusions of Theorem 6.3 [9] and Theorem 6.5 hold for this sequence. In particular, (1) and (2) of the theorem hold for $R_0(0) \rightarrow R_0 \rightarrow S_0$ and $p(\hat{S}_0)_\infty$, $p(\hat{R}_0)_\infty$ have the desired form.

Suppose that m is a positive integer, $f = (f_1, \ldots, f_m) \in (V^*)^m$. Set $u_i = y_i(0)$, $1 \leq i \leq n$. Set $u_{n+i} = f_i$ for $1 \leq i \leq m$. If $f_i \in V^* \cap K = V$, also set $v_i = f_i$.

By Theorem 4.9 [9] and Theorem 6.1 [9], with the R^*, S^* in the assumptions of Theorem 4.9 [9] set as $R^* = R_0(0)$, $S^* = S_0$, and with the $\{u_i\}$ and $\{v_i\}$ defined as above, and then applying Theorem 6.5 (and Remark 6.6), there exists a commutative diagram

$$\begin{array}{ccccc} R_0(f) & \rightarrow & R_f & \rightarrow & S_f \\ \uparrow & & & & \uparrow \\ R_0(0) & \rightarrow & & & S_0 \end{array}$$

such that the vertical arrows are sequences of monoidal transforms along V^*, (1) and (2) of this theorem hold for

$$R_0(f) \rightarrow R_f \rightarrow S_f$$

and (3)(a) of this theorem holds for

$$\begin{array}{ccc} R_f & \rightarrow & S_f \\ \uparrow & & \uparrow \\ R_0 & \rightarrow & S_0 \end{array}.$$

We have that $\det(D(0,f)) \neq 0$ since (1) holds for S_0 and S_f. Define a partial ordering on $I = \sqcup_{m\in \mathbf{N}_+}(V^*)^m$ by $f \le g$ if $S_f \subset S_g$. We will associate to $0 \in V^*$ the sequence $R_0(0) \to R_0 \to S_0$ constructed in the beginning of the proof. Suppose that $\alpha \le \beta$. We have

$$S_0 \subset S_\alpha \subset S_\beta$$

so the proof of (2) of Theorem 7.2 shows that (3)(a) of this Theorem holds for α, β.

(3)(b) holds since

$$R_\alpha = S_\alpha \cap K \subset S_\beta \cap K = R_\beta.$$

(3)(c) is immediate, since the conclusions of Theorem 6.1 [9] hold. In particular, (11) of Theorem 4.7 [9] holds.

Finally, we will establish (4) of the Theorem. By construction, $V^* = \cup_{j\in I} S_j$. If $f \in V$, we have $f \in S_f \cap K = R_f$, thus $V = \cup_{j\in I} R_j$. By construction, $\cup_{j\in I}\Omega_j = \Gamma^*$, since $\nu^*(f) \in \Omega_f$ for $f \in V^*$. We also have $\cup_{j\in I}\Lambda_j = \Gamma$, since $\nu(f) \in \Lambda_f$ for $f \in V$. I is a directed set as shown in the proof of Theorem 7.2. □

Theorem 8.2. *Let assumptions be as in Theorem 8.1. There exists a partially ordered set I and algebraic regular local rings $\{S_i \mid i \in I\}$ with quotient field K^* which are dominated by V^* and algebraic local rings with toric singularities $\{R_i \mid i \in I\}$ such that*

(1)
$$V^* = \lim_{\longrightarrow} S_i = \cup_{i\in I} S_i, V = \lim_{\longrightarrow} R_i = \cup_{i\in I} R_i$$
and each S_i is a localization at a maximal ideal of the integral closure of R_i in K^*.

(2) *$\hat{S}_j/p(\hat{S}_j)_\infty$ are regular local rings for all j and*
$$T^* = \lim_{\longrightarrow} \hat{S}_j/p(\hat{S}_j)_\infty$$
is a Henselian valuation ring such that $(T^*, Q(T^*))$ is an immediate extension of (V^*, K^*).

(3) *$\hat{R}_j/p(\hat{R}_j)_\infty$ has normal toric singularities for all j and*
$$T = \lim_{\longrightarrow} \hat{R}_j/p(\hat{R}_j)_\infty$$
is a Henselian valuation ring such that $(T, Q(T))$ is an immediate extension of (V, K).

(4) *Further suppose that $k = V^*/m_{V^*}$ is algebraically closed (of characteristic zero). Then the action of Γ^*/Γ on $\hat{S}_j$ by k-algebra isomorphisms extends to an action of Γ^*/Γ on $\hat{S}_j/p(\hat{S}_j)_\infty$, and an action of Γ^*/Γ on T^* such that $\hat{R}_j/p(\hat{R}_j)_\infty \cong (\hat{S}_j/p(\hat{S}_j)_\infty)^{\Gamma^*/\Gamma}$ and $T = (T^*)^{\Gamma^*/\Gamma}$.*

Proof. The theorem follows from Theorems 8.1, 7.3 and 7.4. The fact that $p(\hat{S}_j)_\infty$ is fixed by Γ^*/Γ follows from (2) of Theorem 8.1. □

Now the proof of Theorem 1.3 follows from Theorem 8.2. The proof that $Q(U^*)$ is Galois over $Q(U)$ is as in Theorem 1.2.

References

[1] Abhyankar, S., Local uniformization on algebraic surfaces over ground fields of characteristic $p \neq 0$, Annals of Math, 63 (1956), 491-526.

[2] Abhyankar, S., Simultaneous resolution for algebraic surfaces, Amer. J. Math, 78 (1956), 761-790.

[3] Abhyankar, S., Ramification theoretic methods in algebraic geometry, Annals of Math. Studies 43, Princeton University Press, 1959.

[4] Abhyankar, S., Resolution of Singularities of embedded algebraic surfaces, Academic Press, New York, 1966.
[5] Abhyankar, S., Resolution of singularities and modular Galois theory, Bulletin of the AMS, 38, 131-171.
[6] Cutkosky, S.D., Local factorization and monomialization of morphisms, Asterisque, 1999.
[7] Cutkosky, S.D., Simultaneous resolution of singularities, Proc. American Math. Soc. 128, (2000), 1905-1910.
[8] Cutkosky, S.D., Local Monomialization of Transcendental Extensions, preprint 2003, www.math.missouri.edu/~cutkosky
[9] Cutkosky, S.D. and Piltant, O., Ramification of valuations, to appear in Advances in Math., www.math.missouri.edu/~cutkosky
[10] Endler, O., Valuation Theory, Springer Verlag, New York, Heidelberg, Berlin, 1972.
[11] Grothendieck A. and Dieudonné J., Eléments de Géometrie Algébrique IV, Étude locale des schémes et des morphisms de schémas (seconde partie), Publ. Math. IHES 24 (1965).
[12] Hartshorne, R., Algebraic Geometry, Springer-Verlag, 1977.
[13] Hironaka, H., Resolution of singularities of an algebraic variety over a field of characteristic zero, Annals of Math, 79 (1964), 109-326.
[14] Heinzer, W. and Sally, J., Extensions of valuations to the completion of a local domain, Journal of Pure and Applied Algebra 71 (1991), 175-185.
[15] Hungerford, T. , Algebra, Springer Verlag, New York, Heidelberg, Berlin (1974).
[16] Kaplansky, I., Maximal fields with valuations, Duke J. 9 (1942), 303-321.
[17] Krull, W., Galoissche Theorie bewerteter Körper, Sitzungsberichte der Bayerischen Akademie der Wissenschaften, München (1930), 225-238.
[18] Krull, W., Allgemeine Bewertungstheorie, J. reine angew. Math. 167 (1931), 160-196.
[19] Kuhlmann, F.-V., Pank, M. and Roquette, P., Immediate and purely wild extensions of valued fields, Manuscripta Math. 55 (1986), 39-67.
[20] Kuhlmann, F.-V., Valuation theoretic and model theoretic aspects of local uniformization, in Resolution of Singularities, edited by H. Hauser, J. Lipman, F. Oort, A. Quirós, Springer-Verlag, 2000.
[21] Schilling, O.F.G., The theory of valuations, Mathematical Surveys IV, AMS (1950).
[22] Teissier, B., Valuations, Deformations and Toric Geometry, Proceedings of the Saskatoon Conference and Workshop on Valuation Theory, Volume II, F-V Kuhlmann, S. Kuhlmann, M. Marshall editors, Fields Inst. Comm., to appear.
[23] Zariski, O., Local uniformization of algebraic varieties, Math. Ann. 41 (1940), 852-896.
[24] Zariski, O. and Samuel, P., Commutative Algebra 2, Van Nostrand, Princeton, 1960.

Contemporary Mathematics
Volume **386**, 2005

A New Construction of Maximal Curves

Abigail Ebin, Deborah Hur, Aaron Katz, and Vladislav Shchogolev

Communicated by Rachel Pries

ABSTRACT. Fuhrmann, Garcia, and Torres [**FGT97**] proved there is a unique maximal curve of genus $(q-1)^2/4$ defined over $\mathbb{F}_{q^2}$. We give a new construction of this curve as a pullback of the natural action of $\mathrm{PSL}_2(\mathbb{F}_q)$ on the projective line by a Kummer cover.

CONTENTS

1. Introduction and Preliminaries

1.1. Introduction. Maximal curves are of central importance in the development of algebraic coding theory. The correspondence between curves with many points over a finite field and effective error correcting codes is well established. Tsfasman, Vlăduţ, and Zink in 1982 [**TVZ82**] exploited this correspondence to find a sequence of efficient codes which correct many errors. Here we present a new construction of the maximal curve over $\mathbb{F}_{q^2}$ for each odd prime power q studied by Fuhrmann, Garcia, and Torres. Our construction of the curve involves normalizing the fibre product of a Kummer cover and the cover arising from the natural action of $\mathrm{PSL}_2(\mathbb{F}_q)$ on the projective line $\mathbb{P}^1_{\mathbb{F}_q}$. We develop the necessary machinery in depth in Section 2.

1991 *Mathematics Subject Classification.* Primary 14G50; Secondary 11G20, 11T71, 14H30.
Key words and phrases. maximal curve, coding theory, Galois.
The authors were supported by an undergraduate research project in mathematics during summer 2002 at Columbia University sponsored by NSF VIGRE grant DMS-98-10750.

We would like to thank the NSF for their generous support of this VIGRE undergraduate research project during the summer of 2002. Also, the guidance and help of Professor Jeff Achter, Professor Rachel Pries, Sebastian Casalaina-Martin, and Morgan Sherman has been invaluable.

1.2. Coding Theory. We start with the finite field $\mathbb{F}_q$ which will be our *alphabet.* The number of elements q is a power of a prime p. For the purposes of this paper, we will consider only finite fields of characteristic $p > 2$. Any *message word* we wish to encode is an element of $\mathbb{F}_q^j$ for some j. The message word is encoded into a *codeword* in $\mathbb{F}_q^n$ for some $n > j$.

DEFINITION 1.1. For any n, a *code C of length n* over $\mathbb{F}_q$, is a subset of $\mathbb{F}_q^n$.

A code C is called *linear* if C is a vector subspace of $\mathbb{F}_q^n$. The *dimension* k of a linear code C is the dimension of C over $\mathbb{F}_q$. If $x, y \in C$ are codewords, we define the *Hamming distance* $d_H(x, y) = \#\{i : x_i \neq y_i\}$ and the *Hamming weight* $wt(x) = d_H(0, x)$, which are essential to our understanding of what makes a good code. We define the *minumum distance* $d = \min\{d_H(x, y) : x, y \in C, x \neq y\}$.

PROPOSITION 1.2. *The Hamming distance d_H is a metric on C.*

PROOF. That d_H is symmetric is obvious. Also, $d_H(x, y)$ must be positive for $x \neq y$. Furthermore, for x, y, $z \in C$, we see that if x and z differ in a coordinate, then either x and y differ there or y and z differ there or both. Thus $d_H(x, z) \leq d_H(x, y) + d_H(y, z)$. □

PROPOSITION 1.3. *For a linear code, d equals the minimum weight wt_{min}.*

PROOF. Since every weight is also a distance, we see $d \leq wt_{min}$. We can choose two code words x and y where $d_H(x, y) = d$. Since C is linear, $x - y$ is also a code word and $wt(x - y) = d$. So, $wt_{min} \leq d$. Thus we conclude, $d = wt_{min}$. □

The minimum distance is significant because the maximum number of errors a code with minimum distance d can correct is $\lfloor \frac{d-1}{2} \rfloor$. Considering the distance as a metric, this is clear. If the number of errors is greater than $\frac{d-1}{2}$, then there is either not a unique closest code word or the closest code word is not the one intended.

Our goal is to find a code that has both a high information rate, $R = \frac{k}{n}$, and a high relative distance, $\delta = \frac{d}{n}$. This will allow for efficient coding and accurate error correcting.

1.3. Algebraic Geometry and Codes. We see an interesting application of algebraic geometry to coding theory. First, here are some preliminary definitions.

Let X be a nonsingular projective curve defined over the field $\mathbb{F}_q$.

DEFINITION 1.4. Let D be a divisor on X. The space of rational functions associated to D is

$$L(D) := \{f \in \mathbb{F}_q(X) \colon \operatorname{div}(f) + D \geq 0\} \cup \{0\}.$$

Goppa's construction of algebraic-geometric codes is as follows:

DEFINITION 1.5. Let $\mathcal{P} = \{P_1, P_2, \ldots, P_n\} \subset X(\mathbb{F}_q)$ with $P_i \neq P_j$ for $i \neq j$. Let D be a divisor on X such that $\operatorname{supp}(D) \cap \mathcal{P} = \emptyset$. Then the algebraic geometric code associated to $X, \mathcal{P},$ and D is

$$C(X, \mathcal{P}, D) := \{(f(P_1), f(P_2), \ldots, f(P_n)) \colon f \in L(D)\} \subset \mathbb{F}_q^n.$$

Recall that a good code is both efficient (high information rate) and corrects many errors (large minimum distance). In terms of Goppa codes, with a little argument, these criteria will be satisfied for a curve with many rational points relative to its genus.

We state the following theorem and refer the reader to [**Wei48**] for the proof. It is valuable to find curves whose number of points is as close as possible to the following upper bound:

THEOREM 1.6 (Hasse-Weil Bound). *Let X be a nonsingular projective curve of genus g over the field $\mathbb{F}_q$ and set $N = \#X(\mathbb{F}_q)$. Then*

$$|N - (q+1)| \leq 2g\sqrt{q}.$$

A curve with exactly $q + 1 + 2g\sqrt{q}$ points over $\mathbb{F}_q$ is called *maximal.*

2. Mathematical Ingredients

2.1. The Kummer Cover. Let k be an algebraically closed field of characteristic p. For m relatively prime to p, consider the smooth projective curve Y in $\mathbb{P}^1_k \times \mathbb{P}^1_k$ given by the equation $f(x,y) = y^m - x$. The ring extension $k[x] \hookrightarrow k[x,y]/(f)$ defines a cover $g : Y \longrightarrow \mathbb{P}^1_k$ sending $(x,y) \mapsto x$, whose Galois group G we now compute.

We first consider the automorphism $\sigma(y) = \zeta y$, where ζ is a primitive m^{th} root of unity in k. Since the degree of the extension is m and the order of σ is m, we conclude that the extension is Galois, and the Galois group associated to g is $\mathbb{Z}/m\mathbb{Z}$, generated by the automorphism σ.

To find the branch and ramification points in the (x,y)-affine patch of the curve, we look at which points of the curve are fixed by a nontrivial element τ in the Galois group G. Since σ generates G, we may write $\tau(y) = \sigma^i(y) = \zeta^i y$ for some $i \in \mathbb{Z}$, $m \nmid i$. We see that $y = \zeta^i y$ if and only if $y = 0$. So g is ramified at the point 0_Y where $x = y = 0$. To consider $\infty_{\mathbb{P}^1}$, we use a change of coordinates, $\bar{x} = 1/x$ and $\bar{y} = 1/y$. We similarly conclude that in the $(\bar{x},\bar{y})$-patch of the curve, $\bar{y} = \zeta^i \bar{y}$ if and only if $\bar{y} = 0$. Thus the two ramified points are $\mathbf{0}_Y$ and ∞_Y, which lie above the branch points $\mathbf{0}_{\mathbb{P}^1}$ and $\infty_{\mathbb{P}^1}$, respectively.

DEFINITION 2.1. Let $\phi : Y \longrightarrow X$ be a Galois cover of curves with Galois group G. Let $P \in X$ and let $Q \in \phi^{-1}(P)$. The *decomposition group* at Q is $D_Q = \{\sigma \in G : \sigma(Q) = Q\}$. The *inertia group* I_Q at Q is the group of automorphisms in D_Q that induce the identity on the residue field of Y at Q. Over an algebraically closed field, the inertia group equals the decomposition group.

Notice for the cover g that above each of the branch points, there is only one point above it in the fibre. Thus the inertia groups of $\mathbf{0}_Y$ and ∞_Y are both $\mathbb{Z}/m\mathbb{Z}$.

We refer the reader to [**Har77**, IV Corollary 2.4] for the proof of the next theorem.

THEOREM 2.2 (Riemann-Hurwitz Formula). *Let $\phi : Y \longrightarrow X$ be a non-constant separable morphism of non-singular projective curves defined over k. If p does not divide the order of any inertia group and* $\deg(R) = \sum_Q(|I_Q| - 1)$ *is the sum over all ramification points $Q \in Y$, then we have*

$$2g_Y - 2 = \deg(\phi)(2g_X - 2) + \deg(R).$$

Applying the Riemann-Hurwitz formula, we see that:

$$2g_Y - 2 = m(2g_X - 2) + \deg(R) = -2m + 2(m-1).$$

So the genus of Y is 0 and thus $Y \cong \mathbb{P}^1_k$.

2.2. $\mathrm{PSL}_2(\mathbb{F}_q)$. Let $q = p^r$ for some prime p. Let $\mathrm{SL}_2(\mathbb{F}_q)$ be the group of two-by-two matrices whose entries are in $\mathbb{F}_q$ and which have determinant one. Let $H \leqslant \mathrm{SL}_2(\mathbb{F}_q)$ be the normal subgroup $\{I, -I\}$. Let $\mathrm{PSL}_2(\mathbb{F}_q)$ be the quotient group $\mathrm{SL}_2(\mathbb{F}_q)/H$. Now we would like to compute the size of this group.

LEMMA 2.3. *The order of $GL_2(\mathbb{F}_q)$ is $(q^2-1)(q^2-q)$.*

PROOF. A matrix is invertible if and only if its column vectors are linearly independent. Given the matrix $\begin{pmatrix} a & b \\ c & d \end{pmatrix}$, there are q^2-1 choices such that the first column is a nonzero vector. The number of elements in the span of this vector is q, one for each of its scalar multiples. Since the first and second vector must be linearly independent, our choice for the second column is limited to any of the other q^2-q vectors in the space. □

PROPOSITION 2.4. *The order of $PSL_2(\mathbb{F}_q)$ is $q(q^2-1)/2$.*

PROOF. We have

$$|\mathrm{PSL}_2(\mathbb{F}_q)| = (\mathrm{SL}_2(\mathbb{F}_q) : H) = |\mathrm{SL}_2(\mathbb{F}_q)|/2.$$

Furthermore, $\det : \mathrm{GL}_2(\mathbb{F}_q) \longrightarrow \mathbb{F}_q^*$ is a surjective homomorphism with kernel $\mathrm{SL}_2(\mathbb{F}_q)$. Hence $\mathrm{GL}_2(\mathbb{F}_q)/\mathrm{SL}_2(\mathbb{F}_q) \cong \mathbb{F}_q^*$ and $|\mathrm{GL}_2(\mathbb{F}_q)/\mathrm{SL}_2(\mathbb{F}_q)| = q-1$. So,

$$|\mathrm{SL}_2(\mathbb{F}_q)| = \frac{|\mathrm{GL}_2(\mathbb{F}_q)|}{q-1} = \frac{(q^2-1)(q^2-q)}{q-1} = q(q^2-1).$$

Combining these two equations we arrive at the desired result. □

2.3. Subgroups. The set B of lower triangular matrices in $\mathrm{SL}_2(\mathbb{F}_q)$ forms a subgroup under multiplication.

LEMMA 2.5. *The quotient B/H in $PSL_2(\mathbb{F}_q)$ is isomorphic to a semi-direct product $(\mathbb{Z}/p\mathbb{Z})^r \rtimes \mathbb{Z}/\frac{(q-1)}{2}\mathbb{Z}$.*

PROOF. Let t be a generator of the cyclic group $\mathbb{F}_q^*$ and let T in $\mathrm{SL}_2(\mathbb{F}_q)$ be the subgroup generated by $\begin{pmatrix} t & 0 \\ 0 & t^{-1} \end{pmatrix}$. This matrix has order $q-1$. This is because $\begin{pmatrix} t & 0 \\ 0 & t^{-1} \end{pmatrix}^i = \begin{pmatrix} t^i & 0 \\ 0 & t^{-i} \end{pmatrix}$ which is the identity if and only if i is a multiple of $q-1$. So T is a cyclic group of order $q-1$.

Now let $A = \begin{pmatrix} 1 & 0 \\ a & 1 \end{pmatrix}$ where $a \in \mathbb{F}_q$. We have $A^n = \begin{pmatrix} 1 & 0 \\ na & 1 \end{pmatrix}$. Since $\mathbb{F}_q$ has characteristic p, A has order p in $\mathrm{SL}_2(\mathbb{F}_q)$ if $a \neq 0$. Let P be this subset of lower triangular matices with ones on the diagonal. Then there is a group isomorphism $\phi : P \to \langle \mathbb{F}_q, + \rangle$ given by the map $\phi : \begin{pmatrix} 1 & 0 \\ a & 1 \end{pmatrix} \longmapsto a$. To see this, suppose A is as defined above and $B = \begin{pmatrix} 1 & 0 \\ b & 1 \end{pmatrix}$. Then

$$\phi(\mathrm{AB}) = \phi\left(\begin{pmatrix} 1 & 0 \\ a+b & 1 \end{pmatrix}\right) = a + b = \phi(A) + \phi(B).$$

In turn, $\langle \mathbb{F}_q, + \rangle$ is isomorphic to $\langle (\mathbb{Z}/p\mathbb{Z})^r, + \rangle$ if $q = p^r$.

Since P is normalized by T and $P \cap T = \{\mathrm{I}\}$, the group B is isomorphic to a semi-direct product $(\mathbb{Z}/p\mathbb{Z})^r \rtimes \mathbb{Z}/(q-1)\mathbb{Z}$. Also $H \subseteq T$. Thus, $B/H \cong (\mathbb{Z}/p\mathbb{Z})^r \rtimes \mathbb{Z}/\frac{(q-1)}{2}\mathbb{Z}$. □

2.4. Action of $\mathrm{PSL}_2(\mathbb{F}_q)$ on $\mathbb{P}^1$. Next, we examine the action of $\mathrm{PSL}_2(\mathbb{F}_q)$ on the projective line, $\mathbb{P}^1$. First, we must verify that we do in fact get an action of this group on $\mathbb{P}^1$.

PROPOSITION 2.6. *The group $PSL_2(\mathbb{F}_q)$ gives an action on the set $\mathbb{P}^1$.*

PROOF. We define the action of $M = \begin{pmatrix} a & b \\ c & d \end{pmatrix}$ on a choice of coordinates $[x_0 : x_1]$ of a point x in $\mathbb{P}^1$ like this:

$$M(x) = \begin{pmatrix} a & b \\ c & d \end{pmatrix} \begin{pmatrix} x_0 \\ x_1 \end{pmatrix} = \begin{pmatrix} ax_0 + bx_1 \\ cx_0 + dx_1 \end{pmatrix}.$$

First, we show that the action is well defined (i.e., it does not depend on the choice of coordinates for x under the equivalence relation $\sim$ for the projective line).

$$M(tx) = \begin{pmatrix} a & b \\ c & d \end{pmatrix} t \begin{pmatrix} x_0 \\ x_1 \end{pmatrix} = \begin{pmatrix} atx_0 + btx_1 \\ ctx_0 + dtx_1 \end{pmatrix} = t \begin{pmatrix} ax_0 + bx_1 \\ cx_0 + dx_1 \end{pmatrix} \sim M(x).$$

Hence, the action is well-defined. All matrices in $\mathrm{SL}_2(\mathbb{F}_q)$ are invertible, so none can map an element of $\mathbb{P}^1$ to the zero-vector, which is not in $\mathbb{P}^1$. Next we must check that the subgroup $H = \{I, -I\}$ of $\mathrm{SL}_2(\mathbb{F}_q)$ acts trivially on $\mathbb{P}^1$, since we identify elements of H with the identity in $\mathrm{PSL}_2(\mathbb{F}_q)$. This is clear for $I \in H$. For $-I \in H$, we see that $-Ix = -x \sim x$. Thus H acts trivially on $\mathbb{P}^1$ and we get an action of $\mathrm{PSL}_2(\mathbb{F}_q)$ on $\mathbb{P}^1$. □

Now, we look at the orbit of the point $[0 : 1]$ (i.e., the point at ∞) under the action.

PROPOSITION 2.7. *The stabilizer of $[0:1]$ is B/H which has order $q(q-1)/2$. The orbit of $[0:1]$ consists of the $q+1$ points in $\mathbb{P}^1_{\mathbb{F}_q}$.*

PROOF.

We see $\begin{pmatrix} a & b \\ c & d \end{pmatrix} \begin{pmatrix} 0 \\ 1 \end{pmatrix} = \begin{pmatrix} b \\ d \end{pmatrix} \sim \begin{pmatrix} 0 \\ 1 \end{pmatrix}$ if and only if $b = 0$ and $d \neq 0$.

Thus $\mathrm{Stab}\,[0:1] = \left\{ \begin{pmatrix} a & 0 \\ c & a^{-1} \end{pmatrix} : a, c \in \mathbb{F}_q, a \neq 0 \right\} /H$ which is B/H. There are q possibilities for c and $q-1$ possibilities for a, hence there are $q(q-1)$ elements of B. Hence $|B/H| = |B|/2 = q(q-1)/2$.

We see that

$$\begin{pmatrix} a & b \\ c & d \end{pmatrix} \begin{pmatrix} 0 \\ 1 \end{pmatrix} = \begin{pmatrix} b \\ d \end{pmatrix} \sim \begin{pmatrix} 1 \\ d/b \end{pmatrix} \text{ for } b \neq 0 \text{ and } \begin{pmatrix} 0 \\ 1 \end{pmatrix} \text{ for } b = 0.$$

If $b = 0$ we get the point at infinity $[0 : 1]$. If $b \neq 0$, we can get any point of $\mathbb{F}_q$ for d/b. Thus the orbit of the point at infinity is $\mathbb{P}^1_{\mathbb{F}_q}$. □

Now, we look at the other points under the action.

LEMMA 2.8. *The matrix* $M = \begin{pmatrix} a & b \\ c & d \end{pmatrix}$ *fixes* $[1 : x_1]$ *if and only if* $bx_1^2 + (a - d)x_1 - c = 0$.

PROOF. If $M = \begin{pmatrix} a & b \\ c & d \end{pmatrix}$ fixes $[1 : x_1]$, then $\begin{pmatrix} a & b \\ c & d \end{pmatrix}\begin{pmatrix} 1 \\ x_1 \end{pmatrix} = \begin{pmatrix} t \\ tx_1 \end{pmatrix}$ for some $t \in \mathbb{F}_q^*$. So $(a + bx_1)x_1 = tx_1$ and $c + dx_1 = tx_1$. So $(a - d)x_1 + bx_1^2 = c$.

If $bx_1^2 + (a - d)x_1 - c = 0$ then $(c + dx_1)/(a + bx_1) = x_1$. So

$$M\begin{pmatrix} 1 \\ x_1 \end{pmatrix} = \begin{pmatrix} a & b \\ c & d \end{pmatrix}\begin{pmatrix} 1 \\ x_1 \end{pmatrix} = \begin{pmatrix} a + bx_1 \\ c + dx_1 \end{pmatrix} \sim \begin{pmatrix} 1 \\ x_1 \end{pmatrix}.$$

□

PROPOSITION 2.9. *The set* F *of all* x *which are not in the orbit of* $[0 : 1]$ *and which are fixed by some matrix* $M \in PSL_2(\mathbb{F}_q)$ *is* $\mathbb{P}^1_{\mathbb{F}_{q^2}} \setminus \mathbb{P}^1_{\mathbb{F}_q}$.

PROOF. Suppose $x \in F$. Without any loss of generality, we write $x = [1 : x_1]$. If $M = \begin{pmatrix} a & b \\ c & d \end{pmatrix}$ fixes $[1 : x_1]$ then $bx_1^2 + (a - d)x_1 - c = 0$. Hence x_1 satisfies a quadratic equation with coefficients in $\mathbb{F}_q$, so it is contained in $\mathbb{F}_{q^2}$. Recalling that x is not in the orbit of the point at ∞, we see $F \subseteq \mathbb{P}^1_{\mathbb{F}_{q^2}} \setminus \mathbb{P}^1_{\mathbb{F}_q}$.

Conversely, given $x \in \mathbb{P}^1_{\mathbb{F}_{q^2}} \setminus \mathbb{P}^1_{\mathbb{F}_q}$, we show that x is fixed by some matrix, $M \in \mathrm{PSL}_2(\mathbb{F}_q)$. Again without a loss of generality, we can write $x = [1 : x_1]$ where $x_1 \in \mathbb{F}_{q^2} \setminus \mathbb{F}_q$. From the lemma, we want to find a, b, c, d so that $bx_1^2 + (a - d)x_1 - c = 0$. Consider the unique monic irreducible polynomial $x^2 + \alpha x + \beta$, with $\alpha \in \mathbb{F}_q$ and $\beta \in \mathbb{F}_q^*$, which has x_1 as a root. Equating coefficients, we need

$$\begin{aligned} a - d &= b\alpha, \\ \text{and } -c &= \beta b. \end{aligned}$$

These equations give us a matrix M, but its determinant is not necessarily one. Let $M' = (1/\det M)M^2$. Notice that

$$\det(M') = \det\left(\frac{1}{\det M}M^2\right) = \frac{1}{(\det M)^2}(\det M)^2 = 1.$$

If $M \in \mathrm{GL}_2(\mathbb{F}_q)$ fixes x, then M^2 fixes x. Multiplying by a scalar (in this case, $1/(\det M)^2$) also fixes x because of the equivalence relation on the projective line. Thus $M' \in \mathrm{SL}_2(\mathbb{F}_q)$ and fixes x whenever M fixes x.

We only need to find one $M \in \mathrm{SL}_2(\mathbb{F}_q)$ (which fixes x) so that M' is not the identity matrix. If $\alpha \neq 0$, we let $a = 0$ and $b = 1$. From the above equations, we get:

$$M' = \frac{1}{\beta}\begin{pmatrix} 0 & 1 \\ -\beta & -\alpha \end{pmatrix}^2 = \begin{pmatrix} -1 & -\alpha\beta^{-1} \\ \alpha & \alpha^2\beta^{-1} - 1 \end{pmatrix}$$

$$\text{and } \det(M') = 1 - \alpha^2\beta^{-1} + \alpha^2\beta^{-1} = 1.$$

If $\alpha = 0$, then $\beta \neq -1$ since $x_1^2 + \beta$ is irreducible. We let $a = b = 1$. Then $M' = \frac{1}{1+\beta}\begin{pmatrix} 1-\beta & 2 \\ -2\beta & 1-\beta \end{pmatrix}$.

After identifying M' with its coset, we end up with a non-trivial element of $\mathrm{PSL}_2(\mathbb{F}_q)$ which fixes x. Thus $\mathbb{P}^1_{\mathbb{F}_{q^2}} \setminus \mathbb{P}^1_{\mathbb{F}_q} \subseteq F$. □

Now we wish to compute the stabilizer and orbit of an element $x = [1 : x_1]$ of F. Let $x^2 + \alpha x + \beta$ be the irreducible polynomial for x_1 over $\mathbb{F}_q$. Before we can do this we need to define the following group:

$$D_x = \left\{ M = \begin{pmatrix} a & b \\ -b\beta & a - b\alpha \end{pmatrix}, \; a, b \in \mathbb{F}_q \text{ not both zero} \right\}.$$

These are all the matrices that fix a given $x \in F$. To see that $D_x \subset \mathrm{GL}_2(\mathbb{F}_q)$, notice that $\det M = a^2 - ab\alpha + b^2\beta$. If $b = 0$ then $a \neq 0$ so $\det M \neq 0$. If $b \neq 0$ and $\det M = 0$, then $(a/b)^2 - (a/b)\alpha + \beta = 0$. This equation has the same discriminant as $x^2 + \alpha x + \beta$ which is irreducible over $\mathbb{F}_q$. So $(a/b) \notin \mathbb{F}_q$ which gives a contradiction.

To pick out the matrices with determinant one, we construct the homomorphism:

$$f : D_x \longrightarrow \mathbb{F}_q^*$$

where $f(M) = \det M$ for all $M \in D_x$.

PROPOSITION 2.10. *Given $x = [1 : x_1] \in F$ with irreducible polynomial $x^2 + \alpha x + \beta = 0$, $\mathrm{Stab}\,[1 : x_1] = (\ker f)/H$. This group is cyclic of order $(q+1)/2$.*

PROOF. By the fundamental homomorphism theorem,

$$\frac{|D_x|}{|\ker f|} = |\mathrm{Im}\, f| = \frac{q-1}{(\mathbb{F}_q^* : \mathrm{Im}\, f)}.$$

The cardinality of D_x is $q^2 - 1$ since a and b cannot both be zero. Thus,

$$|\ker f| = \frac{|D_x|}{q-1}(\mathbb{F}_q^* : \mathrm{Im}\, f) = (q+1)(\mathbb{F}_q^* : \mathrm{Im}\, f).$$

To get an upper bound for $|\ker f|$, suppose

$$\det M = a^2 - ab\alpha + b^2\beta = 1.$$

Using the quadratic formula gives an equation for a in terms of b. Thus, for every one of the q choices for b, there are at most 2 choices for a. So there are at most $2q$ elements in $\ker f$.

This gives

$$(q+1)(\mathbb{F}_q^* : \mathrm{Im}\, f) = |\ker f| \leq 2q.$$

Since the index is a natural number, by the inequality above, the only choice is $(\mathbb{F}_q^* : \mathrm{Im}\, f) = 1$. Therefore, $|\ker f/H| = (q+1)/2$. Since $\mathrm{Stab}\,[1 : x_1]$ is the inertia group at x and p does not divide $(q+1)/2$, the inertia group is cyclic. □

PROPOSITION 2.11. *The group $PSL_2(\mathbb{F}_q)$ acts transitively on F.*

PROOF. If $x, y \in F$, we need to show there is some $M \in \mathrm{PSL}_2(\mathbb{F}_q)$ such that $Mx = y$. Let $G = \mathrm{PSL}_2(\mathbb{F}_q)$ and let $x \in \mathbb{P}^1_{\mathbb{F}_{q^2}} \setminus \mathbb{P}^1_{\mathbb{F}_q}$. We can write $x = [1 : x_1]$. Let $M = \begin{pmatrix} a & b \\ c & d \end{pmatrix} \in G$. Now we show that $Mx \in F$ by contradiction. Suppose $Mx \in \mathbb{P}^1_{\mathbb{F}_q}$. Notice that $a + bx_1 \neq 0$ since $x_1 \notin \mathbb{F}_q$. So,

$$Mx = \begin{pmatrix} 1 \\ \frac{c+dx_1}{a+bx_1} \end{pmatrix} \Rightarrow \frac{c+dx_1}{a+bx_1} = \gamma \in \mathbb{F}_q.$$

$$\begin{aligned} \text{It follows that } c + dx_1 &= \gamma(a + bx_1) \\ \text{and } (d - \gamma b)x_1 &= \gamma a - c. \end{aligned}$$

Since $a, b, c, d, \gamma \in \mathbb{F}_q$, then $x_1 = (\gamma a - c)(d - \gamma b) \in \mathbb{F}_q$. This is a contradiction. Thus $G \cdot x \subseteq F$. Furthermore, we see

$$|G \cdot x| = \frac{|G|}{|\mathrm{Stab}(x)|} = \frac{q(q^2-1)/2}{(q+1)/2} = q(q-1) = q^2 - q = |F|.$$

Thus, $\mathrm{PSL}_2(\mathbb{F}_q)$ acts transitively on F. □

3. The Construction of the Maximal Curve

3.1. The Construction of the Maximal Curve. Let $f : W \longrightarrow \mathbb{P}^1_k$ be the quotient cover $f : \mathbb{P}^1 \longrightarrow \mathbb{P}^1/\mathrm{PSL}_2(\mathbb{F}_q)$ and $g : Y \longrightarrow \mathbb{P}^1_k$ be the Kummer cover given by the equation $y^m = x$. Let $m = (q+1)/2$. Let Z be the fibre product $W \times_{f,g} Y$:

$$\begin{array}{ccc} Z := W \times_{f,g} Y & \xrightarrow{\pi_2} & Y \\ \Big\downarrow \pi_1 & & \Big\downarrow g \\ W & \xrightarrow{f} & \mathbb{P}^1 \end{array}$$

Recall that f and g have the following properties:

	f		g	
Galois Group	$\mathrm{PSL}_2(\mathbb{F}_q)$		$\mathbb{Z}/m\mathbb{Z}$	
Branch Points	$\mathbf{0}_{\mathbb{P}^1}$	$\infty_{\mathbb{P}^1}$	$\mathbf{0}_{\mathbb{P}^1}$	$\infty_{\mathbb{P}^1}$
Ramification Points	$\mathbb{P}^1_{\mathbb{F}_{q^2}} \setminus \mathbb{P}^1_{\mathbb{F}_q}$	$\mathbb{P}^1_{\mathbb{F}_q}$	$\mathbf{0}_Y$	∞_Y
Number of Ramification Points	$q^2 - q$ points	$q + 1$ points	1	1
Order of Inertia Group	$\frac{q+1}{2}$	$\frac{q(q-1)}{2}$	$\frac{q+1}{2}$	$\frac{q+1}{2}$

By Abhyankar's Lemma, [**Gro71**, Lemma X 3.6], the fiber product Z is singular at the $q^2 - q$ points above $\mathbf{0}$ since f and g are both ramified over $\mathbf{0}_{\mathbb{P}^1}$ with order $(q+1)/2$. Let $\tilde{Z}$ denote the normalization of Z. Let f_1 be the map that takes $\tilde{Z}$ to Y, and let g_1 be the map that takes $\tilde{Z}$ to W. The degree of f_1 is $q(q^2-1)/2$ and g_1 has degree m. The Galois groups associated with the maps f_1 and g_1 are the same as the Galois groups associated with f and g.

Let $F : \tilde{Z} \longrightarrow \mathbb{P}^1_k$. Then $F = f \circ g_1 = g \circ f_1$. Also F has degree $q(q^2-1)m/2$ and Galois group $\mathrm{PSL}_2(\mathbb{F}_q) \times \mathbb{Z}/m\mathbb{Z}$.

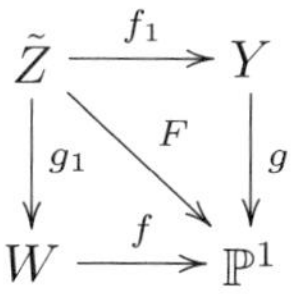

The normalization of Z replaces each singular point with $(q+1)/2$ points. The cover g_1 is then unramified at the $(q+1)(q^2-q)/2$ points in $F^{-1}(\mathbf{0})$.

The order of the ramification of $g^{-1}(\infty)$ is $(q^2-q)/2$, which is relatively prime to $(q+1)/2$. So g_1 is ramified at the points in $F^{-1}(\infty)$ with inertia group $\mathbb{Z}/m\mathbb{Z}$. Thus, there are $q+1$ points in $f_1^{-1}(\infty)$ and one point in $g_1^{-1}(\infty)$, so there are $q+1$ points in $\tilde{Z}$ above ∞, ramified with order $(q+1)/2$.

3.2. Genus of $\tilde{Z}$. We will use the Riemann-Hurwitz formula to determine the genus of $\tilde{Z}$. For a nonconstant separable morphism of projective curves $\varphi : Y \longrightarrow X$ with X of genus g_X, the Riemann-Hurwitz formula 2.2 states that

$$2g_Y - 2 = \deg(\varphi)(2g_X - 2) + \deg R.$$

If $\deg \varphi$ is relatively prime to the characteristic of the base field (which it is here), then $\deg R = \sum_Q (|I_Q| - 1)$, where the sum is over all the ramification points $Q \in Y$.

THEOREM 3.1. *The genus of $\tilde{Z}$ is $(q-1)^2/4$.*

PROOF. We apply Riemann-Hurwitz to the map $g_1 : \tilde{Z} \longrightarrow W$, recalling that W is $\mathbb{P}^1$:

$$2g_{\tilde{Z}} - 2 = \deg(g_1)(2g_{\mathbb{P}^1} - 2) + \deg R.$$

As we have shown before, g_1 is ramified at $q+1$ points and the inertia group for each one is $\mathbb{Z}/m\mathbb{Z}$. Hence $|I_Q| - 1 = m - 1$, and $\deg R = (q+1)(m-1) = (q+1)(q-1)/2$. Furthermore, the genus of the projective line, $\mathbb{P}^1$, is zero. We have

$$g_{\tilde{Z}} = [-(q+1) + \frac{(q+1)}{2}\frac{(q-1)}{2} + 2]/2 = (q-1)^2/4. \qquad \square$$

The computation of the genus can also be made using the map $f_1 : \tilde{Z} \longrightarrow Y$. It is considerably more complicated since p divides the orders of the inertia groups, but it verifies this result.

3.3. Points on $\tilde{Z}$.

THEOREM 3.2. *The curve $\tilde{Z}$ is maximal over $\mathbb{F}_{q^2}$.*

PROOF. First, we look at the $\mathbb{F}_q$-rational points on $\tilde{Z}$. All of the $\mathbb{F}_q$-rational points in W map to ∞ by f and only the point at ∞ itself maps from Y to ∞. In the fibre product $W \times_{f,g} Y$, there are $q+1$ points over ∞, so there are $q+1$ $\mathbb{F}_q$-rational points on $\tilde{Z}$.

Next, we look at the number of $\mathbb{F}_{q^2}$-rational points that are not $\mathbb{F}_q$-rational points. The $\mathbb{F}_{q^2} \setminus \mathbb{F}_q$-rational points of $\tilde{Z}$ map to zero by F. The curve Z is singular at 0, but when we normalize to $\tilde{Z}$, we get m distinct points. Each of these points is $\mathbb{F}_{q^2}$-rational since m divides q^2-1 and so the mth roots of unity are in $\mathbb{F}_{q^2}$. Thus there are $m(q^2-q)$ $\mathbb{F}_{q^2}$-rational points that are not $\mathbb{F}_q$-rational points. Note that

$$m(q^2-q) = (q+1)(q-1)q/2 = (q^3-q^2)/2.$$

Let N be the total number of points over $\mathbb{F}_{q^2}$. We see that N is $(q^3-q^2)/2+q+1$. We compare this to the Hasse-Weil bound:

$$\left|N-(q^2+1)\right| = \left|\frac{q^3-q^2}{2}+q+1-(q^2+1)\right| \leq 2g\sqrt{q^2}.$$

In fact we have $\left|N-(q^2+1)\right| = \left|q(q-1)^2/2\right| = 2gq$. So $\tilde{Z}$ meets the Hasse-Weil bound, i.e., it is maximal over $\mathbb{F}_{q^2}$. □

4. Related Work

Fuhrmann, Garcia, and Torres [**FGT97**] proved there is a unique maximal curve of genus $(q-1)^2/4$ defined over $\mathbb{F}_{q^2}$. We refer the reader to [**FGT97**, Theorem 3.1] for the proof.

THEOREM 4.1 (Fuhrmann, Garcia, Torres). *Let X be a maximal curve over $\mathbb{F}_{q^2}$ of genus $g=(q-1)^2/4$. Then the curve X is $\mathbb{F}_{q^2}$-isomorphic to the one given by*

$$y^q+y=x^{(q+1)/2}.$$

So the curve $\tilde{Z}$ is $\mathbb{F}_{q^2}$-isomorphic to the curve with equation $y^q+y=x^{(q+1)/2}$. We can find another equation for $\tilde{Z}$ using the cover $g_1:\tilde{Z}\to W$. The curve W is $\mathbb{P}^1$ and g_1 is branched at $f^{-1}(\infty_{\mathbb{P}^1})$ (which are the $\mathbb{F}_q$-rational points of $\mathbb{P}^1$) with inertia group $\mathbb{Z}/m\mathbb{Z}$. So $\tilde{Z}$ has equation $z^m=\Pi_{i\in\mathbb{F}_q}(y-i)$ which simplifies to $z^{(q+1)/2}=y^q-y$. These two equations are isomorphic over $\mathbb{F}_{q^2}$ but not over $\mathbb{F}_q$.

Ihara [**Iha81**] and Tsfasman, Vlăduţ, and Zink [**TVZ82**] used modular curves to construct a sequence of codes over $\mathbb{F}_q$ that have a high information rate and a high relative distance. These codes are efficient and have good error-correcting capabilities. It turns out that the curve $\tilde{Z}$ is related to these codes. The cover $f_1:\tilde{Z}\to Y$ has Galois group $\mathrm{PSL}_2(\mathbb{F}_q)$ and is branched at only one point $g^{-1}(\infty_{\mathbb{P}^1})=\infty_Y$. When $q=p$, there is only one cover like this so that $\tilde{Z}$ has genus $(p-1)^2/4$ and it comes from the reduction of the modular curve $X(p)$, [**BW04**, Proposition 5.3].

References

[BW04] I. Bouw and S. Wewers. Stable reduction of modular curves. In *Modular curves and abelian varieties (Barcelona, 2002) Cremona et al eds*, pages 1–22. Progress in Math, vol 224, Birkhäuser, 2004.

[FGT97] R. Fuhrmann, A. Garcia, and F. Torres. On maximal curves. *J. Number Theory*, 67(1):29–51, 1997.

[Gro71] A. Grothendieck. *Revêtements étales et groupe fondamental.* Springer-Verlag, Berlin, 1971. Séminaire de Géométrie Algébrique du Bois Marie 1960–1961 (SGA 1), Dirigé par Alexandre Grothendieck. Augmenté de deux exposés de M. Raynaud, Lecture Notes in Mathematics, Vol. 224.

[Har77] R. Hartshorne. *Algebraic geometry.* Springer-Verlag, New York, 1977. Graduate Texts in Mathematics, No. 52.

[Iha81] Y. Ihara. Some remarks on the number of rational points of algebraic curves over finite fields. *J. Fac. Sci. Univ. Tokyo Sect. IA Math.*, 28(3):721–724 (1982), 1981.

[TVZ82] M. Tsfasman, S. Vlăduţ, and T. Zink. Modular curves, Shimura curves, and Goppa codes, better than Varshamov-Gilbert bound. *Math. Nachr.*, 109:21–28, 1982.

[Wei48] A. Weil. *Sur les courbes algébriques et les variétés qui s'en déduisent.* Actualités Sci. Ind., no. 1041 = Publ. Inst. Math. Univ. Strasbourg **7** (1945). Hermann et Cie., Paris, 1948.

Department of Mathematics, Columbia University, New York, New York 10027
E-mail address: ace2001@columbia.edu
E-mail address: dhur@math.upenn.edu
E-mail address: amk156@columbia.edu
E-mail address: vs299@columbia.edu

Contemporary Mathematics
Volume **386**, 2005

STRONG NON-ISOTRIVIALITY AND RIGIDITY

SÁNDOR J. KOVÁCS

1. Introduction

Throughout the article the groundfield will always be $\mathbb{C}$, the field of complex numbers.

A family is called isotrivial if any two general members are isomorphic. For example the blow up of the projective plane at a single point, considered as a $\mathbb{P}^1$-bundle over $\mathbb{P}^1$, is an isotrivial family since all of its members are isomorphic to the projective line. Since the genus of a curve in a family is constant, and there is only one curve of genus zero, every family of rational curves is isotrivial. However, for higher genus curves, or more generally, higher dimensional varieties, one can find non-isotrivial families.

At the 1962 ICM in Stockholm, Shafarevich conjectured that there exist only finitely many isomorphism classes of non-isotrivial families of smooth projective curves of a given genus over a given base curve. Shafarevich further conjectured that if there is such a family, then the base curve satisfies a certain hyperbolic condition. For definitions and a more precise formulation, see Section §1. The conjecture was confirmed by [Parshin68] for the case of a compact base and by [Arakelov71] in general. It was recently generalized to families of higher dimensional varieties. This generalization however, is not straightforward.

It will be advantageous to work with a compactification of the family. Considering families over a compact base curve B naturally leads to a slightly different view on the problem. Instead of smooth families over a non-compact base, one may work with arbitrary families over a compact base, and consider the locus over which the family is smooth. Clearly these are equivalent situations. An arbitrary family over a compact base gives a smooth family over some open subset and a smooth family over an open curve can be extended to a (not necessarily smooth) family over the projective closure of the curve.

NOTATION 1.1. For a morphism $f : X \to B$, and a point $b \in B$, X_b will denote the scheme theoretic fibre $f^{-1}(b)$.

DEFINITION 1.2. Let B be a smooth projective variety and $\Delta \subseteq B$ a subvariety. A surjective flat morphism with connected fibers will be called a *family*. A family $f\colon X \to B$ is *isotrivial* if $X_a \simeq X_b$ for general points $a, b \in B$. The family $f\colon X \to B$ will be called *admissible with respect to* (B, Δ) if it is not isotrivial and Δ contains the discriminant locus of f, i.e., the map $f\colon X \setminus f^{-1}(\Delta) \to B \setminus \Delta$ is smooth.

Supported in part by NSF Grant DMS-0092165 and a Sloan Research Fellowship.

CONVENTIONS 1.3. Unless explicitly stated otherwise, *admissible* will always mean *admissible with respect to* (B, Δ). Similarly, if $\mathscr{E}$ and $\mathscr{F}$ are $\mathscr{O}_B$-modules, then $\mathscr{E} \otimes_{\mathscr{O}_B} \mathscr{F}$ will be simply denoted by $\mathscr{E} \otimes \mathscr{F}$.

REMARK 1.4. A smooth projective family, all of whose fibers are isomorphic, is locally trivial in the euclidean topology by the Grauert-Fischer theorem [Fischer-Grauert65]. Similarly, an isotrivial family is generically locally trivial in the étale topology. Finally, the automorphism group of a variety of general type is finite [Kobayashi72, III.2.4]. In fact, even the birational automorphism group is finite [Iitaka82, 11.12]. Therefore, if $f : X \to B$ is an isotrivial family of projective varieties of general type, then there exists an étale cover $\tilde{B} \to B \setminus \Delta$ such that the family

$$\tilde{f} \colon \tilde{X} = (X \setminus f^{-1}(\Delta)) \times_{B \setminus \Delta} \tilde{B} \to \tilde{B}$$

is trivial.

Our starting point is the aforementioned conjecture of Shafarevich:

Shafarevich's Conjecture 1.5. *Let B be a smooth projective curve of genus g and $\Delta \subseteq B$ a finite subset. Further let $q \in \mathbb{Z}, q \geq 2$. Then*

(1.5.1) *there exist only finitely many isomorphism classes of admissible families of curves of genus q, and*

(1.5.2) *if $2g - 2 + \#\Delta \leq 0$, then there exist no such families.*

Shafarevich showed a special case of (1.5.2): There exist no smooth families of curves of genus q over $\mathbb{P}^1$. (1.5.1) was confirmed by [Parshin68] for $\Delta = \emptyset$ and by [Arakelov71] in general.

The ultimate goal is to generalize this statement to higher dimensional families. In order to work toward that goal, the statement has to be reformulated following Parshin and Arakelov.

DEFINITION 1.6. A *deformation (over T)* of a family $g \colon Y \to S$ (with the base fixed) is a family $\mathfrak{g} \colon \mathfrak{Y} \to S \times T$, such that for some $t_0 \in T$, $(\mathfrak{Y}_{t_0} \to S \times \{t_0\}) \simeq (Y \to S)$:

$$\begin{array}{ccc} Y \simeq \mathfrak{Y}_{t_0} & \longrightarrow & \mathfrak{Y} \\ {\scriptstyle g}\downarrow & & \downarrow{\scriptstyle \mathfrak{g}} \\ S \simeq S \times \{t_0\} & \longrightarrow & S \times T. \end{array}$$

For simplicity, $S \times \{t\}$ will be denoted by S_t. Two families $Y_1 \to S$ and $Y_2 \to S$ are said to have the same *deformation type* if they can be deformed into each other, i.e., if there exists a connected T and a deformation of families, $\mathfrak{Y} \to S \times T$ such that for some $t_1, t_2 \in T$, $(\mathfrak{Y}_{t_i} \to S_{t_i}) \simeq (Y_i \to S \times T)$ for $i = 1, 2$.

Next consider deformations of admissible families over the base $B \setminus \Delta$. Doing so potentially allows more deformations than over the original base B: it can easily happen that a deformation over $B \setminus \Delta$, that is, a family $\mathfrak{X} \to (B \setminus \Delta) \times T$, cannot be compactified to a (flat) family over $B \times T$, because the compactification may contain fibers of dimension that are higher than expected. This however, will not cause any problems because of the nature of the present inquiry. One wants to argue the opposite way. If one can prove rigidity for a family over $B \setminus \Delta$, than it will automatically apply to any family over B as well.

With regard to the Shafarevich conjecture, Parshin made the following observation. In order to prove that there are only finitely many admissible families, one can try to proceed the following way. Instead of aiming for the general statement immediately, first try to prove that there are only finitely many deformation types. This is expected to be somewhat easier, because there are ways to parametrize deformations, and the fact that there are only finitely many types translates to the parameter space being of finite type. The next step then is to prove that admissible families are rigid, that is, they do not admit non-trivial deformations. Notice that if one proves these statements for families over $B \setminus \Delta$, then they also follow for families over B. Now, if every deformation type contains only one family, and there are only finitely many deformation types, the original statement follows.

The following is the reformulation of the Shafarevich conjecture according to the principle just outlined. This was already used by Parshin and Arakelov.

REFORMULATION 1.7. Let B be a smooth projective curve of genus g and $\Delta \subseteq B$ a finite subset. Further let $q \in \mathbb{Z}, q \geq 2$. Recall the convention from (1.3).

(1.7.1) *Boundedness*: There exist only finitely many deformation types of admissible families, i.e., admissible families of curves of genus q are parametrized by $\mathbb{T}$, a scheme of finite type.

(1.7.2) *Rigidity*: There exist no non-trivial deformations of any admissible family of curves of genus q, in particular, $\dim \mathbb{T} = 0$.

(1.7.3) *Hyperbolicity*: There do not exist admissible families of curves of genus q if $2g - 2 + \#\Delta \leq 0$, i.e., $\mathbb{T} \neq \emptyset \Rightarrow 2g - 2 + \#\Delta > 0$.

Notice that the notion of rigidity here is somewhat different from the one commonly used. A family will be called *rigid* if its deformation space is zero-dimensional. This intention is made precise below.

DEFINITION 1.8. A deformation, $\mathfrak{g}\colon \mathfrak{Y} \to S \times T$, is called *essentially trivial* if there exists an open neighborhood T_0 of $t_0 \in T$ such that for all $t \in T_0$, $(\mathfrak{Y}_t \to S_t) \simeq (Y \to S)$. It is called *trivial* if $\mathfrak{Y} = Y \times_S (S \times T)$ and $\mathfrak{g} = g \times \pi_S$. Notice that if $T = \operatorname{Spec} R$ where R is a local ring, then an essentially trivial family is also trivial.

Arakelov-Parshin rigidity is said to hold for a family $g\colon Y \to S$ if it does not admit a non-trivial deformation over $T = \operatorname{Spec} R$ where R is a DVR. By a slight abuse of terminology, in this case g will be also called simply *rigid*.

The following notion will also be needed.

DEFINITION 1.9. Let $g : Y \to S$ be a family of projective varieties of general type. By [Kollár87, 2.5] there exists an open subset $S_0 \subseteq S$ and a morphism $\nu_S : S_0 \to Z$ such that Y_s and Y_t ,for $s, t \in S_0$, are birational if and only if $\nu_S(s) = \nu_S(t)$. Then the *variation of g in moduli*, denoted by $\operatorname{Var} g$, is defined as the dimension of the image of ν_S, that is, $\operatorname{Var} g := \dim \overline{\nu_S(S)}$. Obviously $0 \leq \operatorname{Var} g \leq \dim S$.

REMARK 1.10. Let $g : Y \to S$ be a family of varieties of general type. Then using Kollár's birational moduli map, ν_S, it is easy to see that a deformation $\mathfrak{g} : \mathfrak{Y} \to S \times T$ of g is essentially trivial if and only if $\operatorname{Var} \mathfrak{g} = \operatorname{Var} g$. In particular, if $\dim T = 1$ and g has maximal variation in moduli, i.e., $\operatorname{Var} g = \dim S$, then $\mathfrak{g}$ is essentially trivial if and only if $\operatorname{Var} \mathfrak{g} < \dim(S \times T)$.

This implies that a family $g : Y \to S$ with $\operatorname{Var} g = \dim S$ is *rigid* if and only if there does not exist a deformation $\mathfrak{g}\colon \mathfrak{Y} \to S \times T$ of g with $\operatorname{Var} \mathfrak{g} = \dim(S \times T)$.

Next an equivalent criterion for a family to be rigid will be given.

Lemma 1.11. *Let $g : Y \to S$ be a family of varieties of general type with* $\operatorname{Var} g = \dim S$. *Let T be a smooth curve. Then g is rigid if and only if all of its deformations over T are essentially trivial. Equivalently, g is rigid if and only if for any deformation $\mathfrak{g}$ over T,* $\operatorname{Var} \mathfrak{g} < \dim(S \times T)$.

PROOF. Let $\mathfrak{g}\colon \mathfrak{Y} \to S \times T$ be a deformation of g over T. Let $t \in T$ and $R = \mathscr{O}_{T,t}$. Clearly, $\mathfrak{g}$ is essentially trivial if and only if its pull-back to $\operatorname{Spec} R$ is trivial. The latter is equivalent to g being rigid. □

REMARK 1.12. As discussed above, (1.7.1) and (1.7.2) together imply (1.5.1) and (1.7.3) is clearly equivalent to (1.5.2).

Recently there has been a flurry of activity regarding higher dimensional generalizations of (1.7.1) and (1.7.3), cf. [Migliorini95], [Zhang97], [Bedulev-Viehweg00], [Oguiso-Viehweg01], [Viehweg-Zuo01,02,03], [Kovács96,97abc,00,02,03ab]. For detailed surveys one can turn to [Viehweg00] and [Kovács03c].

The focus of this note is (1.7.2): Rigidity. It should also be noted, that some of the results here have been independently obtained by [Viehweg-Zuo04], although their main emphasis is somewhat different.

Unfortunately rigidity, as stated, fails for higher dimensional families as the following simple example shows.

EXAMPLE 1.13. Let $Y \to B$ be an arbitrary non-isotrivial family of curves of genus ≥ 2, and C a smooth projective curve of genus ≥ 2. Then $f\colon X = Y \times C \to B$ is an admissible family, and a deformation of C gives a deformation of f. Therefore (1.7.2) fails as stated.

However, this example gives the feeling of cheating. The family defined here is indeed a non-isotrivial family, but only in one direction. It contains, as a term of a fibered product, an isotrivial family, and that's what makes rigidity fail. The goal of this article is to give a definition of a stronger non-isotriviality notion, that does not allow this to happen and prove that for that notion, rigidity holds. The main result is the following. For the definition of strong non-isotriviality, see (4.7).

Theorem 1.14. *Let f be an admissible family of projective varieties of general type. If f is strongly non-isotrivial over B, then Arakelov-Parshin rigidity holds for f.*

ACKNOWLEDGEMENT. I would like to thank the referee for a very careful reading of this article and pointing out many little details that could improve the presentation.

2. Setup

Let $f : X \to B$ be a projective family where X and B are smooth varieties. Then the short exact sequence,

$$0 \to T_{X/B} \to T_X \to f^*T_B \to 0,$$

induces an $\mathscr{O}_B$-homomorphism,

$$\rho_f : T_B \to R^1 f_* T_{X/B},$$

the *Kodaira-Spencer map* of f. It is well-known that f is isotrivial if and only if ρ_f is injective.

GOALS AND EXPECTATIONS 2.1. One would like to find a condition, call it *strong non-isotriviality*, that strengthens the notion of "non-isotriviality" so that rigidity holds for every strongly non-isotrivial family. Naturally, the desired condition should be reasonably weak but still have this property.

As in the case of non-isotriviality, one expects that strong non-isotriviality will only depend on the general behavior of the family. For $f : X \to B$, and a non-empty open subset $B_0 \subseteq B$, let $f_0 : X_0 = f^{-1}(B_0) \to B_0$. Then one expects that f is strongly non-isotrivial if and only if so is f_0. Notice that if rigidity holds for f_0, then it also holds for f.

REMARK 2.2. The expectation that strong non-isotriviality should be invariant under restriction to a smaller base and the resulting guiding philosophy means that one is seeking a criterion for rigidity in a local sense. It is also worth studying rigidity with the base fixed, but that will not be addressed here.

3. Case study: Rigidity for products

EXAMPLE 3.1. Let B be a smooth projective curve, and let $g : Y \to B$ and $h : Z \to B$ be two families of smooth projective curves of genus at least two. Finally, let $X = Y \times_B Z$.

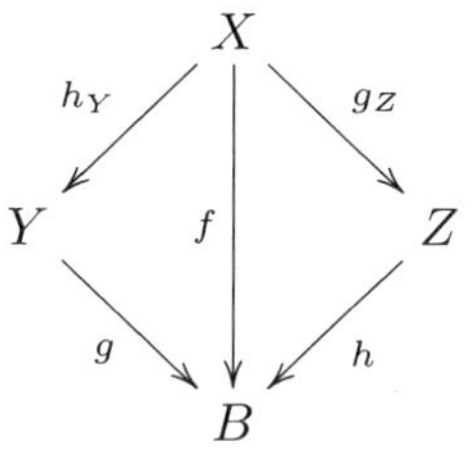

Observe that if either g or h is isotrivial, then after an étale base change it becomes trivial and hence rigidity fails for the pull-back of f. Because of this, f should not be called strongly non-isotrivial if either g or h is isotrivial.

CLAIM 3.2. Using the notation from (3.1),

$$H^1(X, T_{X/B}) \simeq H^1(Y, T_{Y/B}) \oplus H^1(Z, T_{Z/B}).$$

PROOF. $T_{X/B} \simeq h_Y^* T_{Y/B} \oplus g_Z^* T_{Z/B}$ implies that

$$H^1(X, T_{X/B}) \simeq H^1(X, h_Y^* T_{Y/B}) \oplus H^1(X, g_Z^* T_{Z/B}).$$

Since $X = Y \times_B Z$, $R^1(h_Y)_* \mathscr{O}_X \simeq H^1(Z, \mathscr{O}_Z) \otimes \mathscr{O}_Y$, hence

$$H^0(Y, R^1(h_Y)_* \mathscr{O}_X \otimes T_{Y/B}) \simeq H^1(Z, \mathscr{O}_Z) \otimes H^0(Y, \omega_{Y/B}^{-1}) = 0.$$

Then by the Leray spectral sequence, $H^1(Y, T_{Y/B}) \simeq H^1(X, h_Y^* T_{Y/B})$, and similarly $H^1(Z, T_{Z/B}) \simeq H^1(X, g_Z^* T_{Z/B})$. □

Proposition 3.3. *Still using the notation from (3.1), if g and h are both non-isotrivial, then $H^1(X, T_{X/B}) = 0$. In particular rigidity holds for f.*

PROOF. If g is non-isotrivial, then $\omega_{Y/B}$ is ample by [Kovács96, 2.16], cf. [Kovács03, 11.15]. Then by Kodaira vanishing

$$H^1(Y, T_{Y/B}) = H^1(Y, \omega_{Y/B}^{-1}) = 0.$$

Similarly $H^1(Z, T_{Z/B}) = 0$, and hence by (3.2) $H^1(X, T_{X/B}) = 0$. □

CONCLUSION 3.4. For a fibered product of smooth projective families of curves the product family, $f = g \times_B h$ will be called *strongly non-isotrivial* if both g and h are non-isotrivial. It follows that for such an f rigidity holds. Thus one should define 'strong non-isotriviality' in such a way that for a product it coincides with this definition.

4. Iterated Kodaira-Spencer maps

Start with the previous situation, that is, $g : Y \to B$ and $h : Z \to B$ are families of curves of genus at least two and $f = g \times_B h : X = Y \times_B Z \to B$ their fibered product.

Consider the Kodaira-Spencer maps associated with g and h: $\rho_g : T_B \to R^1 g_* T_{Y/B}$, $\rho_h : T_B \to R^1 h_* T_{Z/B}$, and their tensor product:

$$\rho_g \otimes \rho_h : T_B^{\otimes 2} \to R^1 g_* T_{Y/B} \otimes R^1 h_* T_{Z/B}.$$

From (3.3) one obtains the following.

Corollary 4.1. *If $\rho_g \otimes \rho_h \neq 0$ then rigidity holds for f.*

4.2 Iterated Kodaira-Spencer maps: products. Let $\wedge^m T_X$ be denoted by T_X^m, and $\wedge^m T_{X/B}$ by $T_{X/B}^m$. Since $g_* T_{Y/B} = 0$ and $h_* T_{Z/B} = 0$ the Künneth formula implies that

$$R^1 g_* T_{Y/B} \otimes R^1 h_* T_{Z/B} \simeq R^2 f_* T_{X/B}^2.$$

Consider the short exact sequence

$$0 \to T_{X/B} \to T_X \to f^* T_B \to 0.$$

As above this induces the Kodaira-Spencer map of f: $\rho_f : T_B \to R^1 f_* T_{X/B}$. Let us define

$$\rho_f^{(1)} = \rho_f \otimes \mathrm{id}_{T_B} : T_B^{\otimes 2} \to R^1 f_* T_{X/B} \otimes T_B.$$

By taking exterior powers one obtains the short exact sequence,

$$0 \to T_{X/B}^2 \to T_X^2 \to T_{X/B} \otimes f^* T_B \to 0,$$

which in turn induces the map

$$\rho_f^{(2)} : R^1 f_* T_{X/B} \otimes T_B \to R^2 f_* T_{X/B}^2.$$

Notice that one can compose $\rho_f^{(1)}$ and $\rho_f^{(2)}$ to obtain a map,

$$\rho_f^{(2)} \circ \rho_f^{(1)} : T_B^{\otimes 2} \to R^2 f_* T_{X/B}^2.$$

Observation 4.3. $\rho_f^{(2)} \circ \rho_f^{(1)} = \rho_g \otimes \rho_h$.

This statement is not too hard to prove, but unfortunately I do not know a short proof. By writing out explicitly what each side means one can conclude that they are indeed the same.

Corollary 4.4. *If $\rho_f^{(2)} \circ \rho_f^{(1)} \neq 0$, then rigidity holds for f.*

REMARK 4.5. Notice that this statement no longer makes reference to the product structure. This fact will be important in dealing with more general families.

4.6 Iterated Kodaira-Spencer maps: one-dimensional base. Let $f : X \to B$ be a smooth projective family of varieties of general type of dimension n, B a smooth (not necessarily projective) curve and keep using the notation T_X^m for $\wedge^m T_X$ and $T_{X/B}^m$ for $\wedge^m T_{X/B}$.

Let $1 \leq p \leq n$ and consider the short exact sequence,

$$0 \to T_{X/B}^p \otimes f^* T_B^{\otimes(n-p)} \to T_X^p \otimes f^* T_B^{\otimes(n-p)} \to T_{X/B}^{p-1} \otimes f^* T_B^{\otimes(n-p+1)} \to 0.$$

This induces an edge map,

$$\rho_f^{(p)} : R^{p-1} f_* T_{X/B}^{p-1} \otimes T_B^{\otimes(n-p+1)} \to R^p f_* T_{X/B}^p \otimes T_B^{\otimes(n-p)}.$$

DEFINITION 4.7. Let $\rho_f := \rho_f^{(n)} \circ \rho_f^{(n-1)} \circ \cdots \circ \rho_f^{(1)} : T_B^{\otimes n} \longrightarrow R^n f_* T^n_{X/B}$ and call f *strongly non-isotrivial* if $\rho_f \neq 0$.

EXAMPLE 4.8. Let $Y_i \to B$ be non-isotrivial families of smooth projective curves for $i = 1, \dots, r$. Then $X = Y_1 \times_B \cdots \times_B Y_r \to B$ is strongly non-isotrivial.

REMARK 4.9. Since T_B is a line bundle and $R^n f_* T^n_{X/B}$ is locally free, $\rho_f \neq 0$ if and only if it is injective.

4.10 Iterated Kodaira-Spencer maps: higher-dimensional base. Let $f : X \to B$ be a smooth projective family of varieties of general type of dimension n, B a smooth (not necessarily projective) variety.

For an integer p, $1 \leq p \leq n$, there exists a filtration

$$T_X^p = \mathscr{F}^0 \supseteq \mathscr{F}^1 \supseteq \cdots \supseteq \mathscr{F}^p \supseteq \mathscr{F}^{p+1} = 0,$$

such that

$$\mathscr{F}^i \Big/ \mathscr{F}^{i+1} \simeq T^i_{X/B} \otimes f^* T_B^{p-i}.$$

In particular,

$$\mathscr{F}^p \simeq T^p_{X/B}$$

and

$$\mathscr{F}^{p-1} / \mathscr{F}^p \simeq T^{p-1}_{X/B} \otimes f^* T_B.$$

Therefore

$$0 \to T^p_{X/B} \otimes f^* T_B^{\otimes(n-p)} \to \mathscr{F}^{p-1} \otimes f^* T_B^{\otimes(n-p)} \to T^{p-1}_{X/B} \otimes f^* T_B^{\otimes(n-p+1)} \to 0$$

induces a map

$$\rho_f^{(p)} : R^{p-1} f_* T^{p-1}_{X/B} \otimes T_B^{\otimes(n-p+1)} \to R^p f_* T^p_{X/B} \otimes T_B^{\otimes(n-p)}.$$

DEFINITION 4.11. Let $\rho_f := \rho_f^{(n)} \circ \rho_f^{(n-1)} \circ \cdots \circ \rho_f^{(1)} : T_B^{\otimes n} \longrightarrow R^n f_* T^n_{X/B}$ and call f *strongly non-isotrivial over* B if ρ_f is injective.

EXAMPLE 4.12. Let $Y_i \to B$ be non-isotrivial families of smooth projective curves for $i = 1, \dots, r$. Then $X = Y_1 \times_B \cdots \times_B Y_r \to B$ is strongly non-isotrivial over B.

REMARK 4.13. One could consider various refinements:

- Consider maps for which the composition of fewer $\rho^{(p)}$'s is injective or non-zero. This is important in particular to study moduli spaces of varieties that are products with one rigid term.
- Combine this condition with Var f, the variation of f in moduli.

Theorem 4.14. *Let $f : X \to B$ be a smooth projective family of varieties of general type, B a smooth variety. If f is strongly non-isotrivial over B, then rigidity holds for f.*

PROOF. Let $\mathfrak{f} : \mathfrak{X} \to B \times T$ be a deformation of f, and $t_0 \in T$ such that $\mathfrak{f}_{t_0} : \mathfrak{X}_{t_0} \to B_{t_0}$ is isomorphic to $f : X \to B$. Here $B_{t_0} = B \times \{t_0\}$. Notice that without loss of generality one may assume that $\dim T = 1$

Using Grothendieck-Serre duality it is easy to see that $\rho_f : T_B^{\otimes n} \to R^n f_* T^n_{X/B}$ being injective implies that $f_* \omega_{X/B}^{\otimes 2} \to \Omega_B^{\otimes n}$ is generically surjective. From that and Nakayama's lemma one obtains that

$$\mathfrak{f}_* \omega^{\otimes 2}_{\mathfrak{X}/B \times T} \to \Omega^{\otimes n}_{B \times T/T} \tag{4.14.1}$$

is generically surjective in a neighborhood of B_{t_0}.

By (1.11), rigidity fails for f if and only if there exists an $\mathfrak{f}$ such that $\operatorname{Var}\mathfrak{f} = \dim(B\times T)$. Assume that $\operatorname{Var}\mathfrak{f} = \dim(B \times T)$ and try to derive a contradiction.

By [Kollár87, Theorem p.363] and [Esnault-Viehweg91, 0.1] $\mathfrak{f}_*\omega^{\otimes 2}_{\mathfrak{X}/B\times T}$ is big and then by (4.14.1) so is $\Omega^{\otimes n}_{B\times T/T}$. However, $\Omega^{\otimes n}_{B\times T/T} = \pi_B^*\Omega_B^{\otimes n}$, which means that it cannot be big. This is a contradiction, so the statement is proven. □

Corollary 4.15. *Theorem 1.14 follows.*

PROOF. If $f : X \to B$ is admissible with respect to (B, Δ), then let $B^\circ := B \setminus \Delta$, $X^\circ := f^{-1}(B^\circ)$, and $f^\circ := f|_{X^\circ}$. Next apply (4.14) for $f^\circ : X^\circ \to B^\circ$. Finally, notice that if f° is rigid, then so is f. □

References

[Arakelov71] S. ARAKELOV: *Families of algebraic curves with fixed degeneracies*, Izv. A. N. SSSR, **35** (1971), 1269–1293.

[Beauville81] A. BEAUVILLE: *Le nombre minimal de fibers singulières d'une courbe stable sur* $\mathbb{P}^1$, Astérisque, **86** (1981), 97–108.

[Bedulev-Viehweg00] E. BEDULEV AND E. VIEHWEG: *On the Shafarevich conjecture for surfaces of general type over function fields*, Invent. Math., **139** (2000), 603–615.

[Esnault-Viehweg91] H. ESNAULT AND E. VIEHWEG: *Ample sheaves on moduli schemes*, in Algebraic geometry and analytic geometry (Tokyo, 1990), 53–80, ICM-90 Satell. Conf. Proc., Springer, Tokyo, 1991.

[Faltings83] G. FALTINGS: *Arakelov's Theorem for abelian varieties*,Invent. Math., **73** (1983), 337–348.

[Fischer-Grauert65] W. FISCHER AND H. GRAUERT: *Lokal-triviale Familien kompakter komplexer Mannigfaltigkeiten* Nachr. Akad. Wiss. Göttingen Math.-Phys. Kl. II (1965) 89–94.

[Kobayashi72] S. KOBAYASHI: *Transformation groups in differential geometry*. Classics in Mathematics. Springer-Verlag, Berlin, 1995. Reprint of the 1972 edition.

[Iitaka82] S. IITAKA: *Algebraic geometry*, volume 76 of *Graduate Texts in Mathematics*. Springer-Verlag, New York, 1982. An introduction to birational geometry of algebraic varieties, North-Holland Mathematical Library, 24.

[Kollár87] J. KOLLÁR: *Subadditivity of the Kodaira dimension: Fibers of general type*, in: *Algebraic Geometry* (Sendai, 1985), (T. Oda, ed.), Advanced Studies in Pure Math., vol. 10, Kinokuniya-North Holland (1987), pp. 361–398.

[Kovács96] S. J. KOVÁCS: *Smooth families over rational and elliptic curves*, J. Algebraic Geom., **5** (1996), 369–385.

[Kovács97a] S. J. KOVÁCS: *On the minimal number of singular fibers in a family of surfaces of general type*, J. reine angew. Math., **487** (1997), 171–177.

[Kovács97b] S. J. KOVÁCS: *Families over a base with a birationally nef tangent bundle*,Math. Ann., **308** (1997), 347–359.

[Kovács97c] S. J. KOVÁCS: *Relative De Rham complexes for non-smooth morphisms*, in: Birational Algebraic Geometry (Y. Kawamata, V. Shokurov eds.), Contemp. Math., vol. 207 (1997), pp. 89–100.

[Kovács00] S. J. KOVÁCS: *Algebraic hyperbolicity of fine moduli spaces*, J. Algebraic Geom., **9** (2000), 165–174.

[Kovács02] S. J. KOVÁCS: *Logarithmic vanishing theorems and Arakelov-Parshin boundedness for singular varieties*, Compositio Math., **131** (2002), 291–317.

[Kovács03a] S. J. KOVÁCS: *Vanishing theorems, boundedness and hyperbolicity over higher dimensional bases*, Proc. Amer. Math. Soc. **131** (2003), no. 11, 3353–3364.

[Kovács03b] S. J. KOVÁCS: *Viehweg's conjecture for families over* $\mathbb{P}^n$, Special issue in honor of Steven L. Kleiman. Comm. Algebra **31** (2003), no. 8, 3983–3991.

[Kovács03c] S. J. KOVÁCS: *Families of varieties of general type: the Shafarevich conjecture and related problems*, Higher dimensional varieties and rational points (Budapest, 2001), 133–167, Bolyai Soc. Math. Stud., 12, Springer, Berlin, 2003.

[Migliorini95] L. MIGLIORINI: *A smooth family of minimal surfaces of general type over a curve of genus at most one is trivial*, J. Algebraic Geom., **4** (1995), 353–361.

[Oguiso-Viehweg01] K. OGUISO AND E. VIEHWEG: *On the isotriviality of families of elliptic surfaces*, J. Algebraic Geom., **10** (2001), 569–598.

[Parshin68] A. PARSHIN: *Algebraic curves over function fields*,Izv. A.N. SSSR, **32** (1968), 1145–1170.

[Viehweg00] E. VIEHWEG: *Positivity of direct image sheaves and applications to families of higher dimensional manifolds* (2000) Lecture notes, "School on Vanishing Theorems and Effective Results in Algebraic Geometry", ICTP, Trieste.

[Viehweg-Zuo01] E. VIEHWEG AND K. ZUO: *On the isotriviality of families of projective manifolds over curves*, J. Algebraic Geom., **10** (2001), 781–799.

[Viehweg-Zuo02] E. VIEHWEG AND K. ZUO: *Base spaces of non-isotrivial families of smooth minimal models*, Complex geometry (Göttingen, 2000), 279–328, Springer, Berlin, 2002.

[Viehweg-Zuo03] E. VIEHWEG AND K. ZUO: *On the Brody hyperbolicity of moduli spaces for canonically polarized manifolds*, Duke Math. J. **118** (2003), no. 1, 103–150.

[Viehweg-Zuo04] E. VIEHWEG AND K. ZUO, *Complex multiplication, Griffiths-Yukawa couplings, and rigidity for families of hypersurfaces*, preprint

[Zhang97] Q. ZHANG: *Global holomorphic one-forms on projective manifolds with ample canonical bundles*, J. Algebraic Geom. **6** (1997), 777–787.

UNIVERSITY OF WASHINGTON, DEPARTMENT OF MATHEMATICS, SEATTLE, WA 98195-4350, USA
E-mail address: kovacs@math.washington.edu
URL: http://www.math.washington.edu/~kovacs

Contemporary Mathematics
Volume **386**, 2005

Spectral sequences associated to morphisms of locally free sheaves

Sándor J. Kovács

April 24, 2005

Let

$$0 \longrightarrow \mathcal{F}' \longrightarrow \mathcal{F} \longrightarrow \mathcal{F}'' \longrightarrow 0$$

be a short exact sequence of locally free sheaves. Then for any $r \geq 0$ there is a finite filtration of $\bigwedge^r \mathcal{F}$,

$$\bigwedge^r \mathcal{F} = F^0 \supset F^1 \supset \cdots \supset F^r \supset F^{r+1} = 0,$$

with quotients

$$F^p/F^{p+1} \simeq \bigwedge^p \mathcal{F}' \otimes \bigwedge^{r-p} \mathcal{F}''$$

for each p.

This filtration in turn leads to the spectral sequence,

$$E_1^{p,q} = H^{p+q}(X, \bigwedge^p \mathcal{F}' \otimes \bigwedge^{r-p} \mathcal{F}'') \Rightarrow H^{p+q}(X, \bigwedge^r \mathcal{F}).$$

This is a very useful tool, but requires $\mathcal{F}', \mathcal{F}, \mathcal{F}''$ be locally free. The aim of the present article is to give a construction that yields a similar spectral sequence in a more general case. More precisely, the goal is to define natural objects that will play the role of $\bigwedge^{r-p} \mathcal{F}''$ if the cokernel of the morphism $\mathcal{F}' \to \mathcal{F}$ is not locally free. These objects will not be single sheaves anymore, but objects of the derived category of $\mathcal{O}_X$-modules. Nevertheless their functorial and cohomological properties will be strong enough to make applications possible.

This work originated in [**Kovács96**]. The principal goal was to make a cohomological argument work in a more general setting and the path to that goal led through the construction of certain objects whose properties resemble those of the components of the relative De Rham complex of a smooth morphism. In particular

1991 *Mathematics Subject Classification.* Primary 14F05; Secondary 14F17.
Supported in part by NSF Grant DMS-0092165 and a Sloan Research Fellowship.

the analogue of the above spectral sequence was constructed in a special case (cf. [**ibid.**]).

Subsequently I improved the construction and later I realized that similar arguments can be used elsewhere. The constructions needed were very similar, yet somewhat different cf. [**Kovács97b, Kovács02**]. It seems appropriate to give a common generalization, so one would only have to go through the construction once rather then to include it with every application, not to mention the hope that there will be new applications. The present article attempts to provide a unified treatment.

§1 contains some technical material that will be essential later. §§1.1 ought to be known, but I do not know a convenient reference, so some of the statements are proved here. §§1.2 is a simple generalization of the notion of a filtration to the derived categorical setting. (1.2.2) is the central result of this section, it is the analogue of its well-known relative about the spectral sequence associated to a filtration.

In §2 the complexes $\underline{\Omega}^p_{\theta_X}$ are constructed for a morphism θ_X between locally free sheaves. If θ_X is the first morphism of the above short exact sequence and its cokernel is locally free, then $\underline{\Omega}^p_{\theta_X} \simeq \bigwedge^p \mathcal{F}''$. The guiding principles of the construction are functoriality and the desired "filtration". The main result is (2.7).

The results in this article were first developed for their use in [**Kovács97b**] in a more specific form. Later the similar, but still different needs of another project prompted this more general presentation. Hence the applications form an important part.

The first application is a vanishing theorem, proved in §3. This is a generalization of both [**Kovács96, 2.13**] and [**Kovács97a, 1.4**]. The functorial property of $\underline{\Omega}^p_{\theta_X}$ is proved in §4.

§5 gives a brief account of the more explicit applications of the theory. These statements are not proved here, the reader is referred to the cited articles for more details.

Definitions and Notation. Let X be a scheme over T. Then $C(X)$ is the category of complexes of $\mathcal{O}_X$-modules and for $u \in \operatorname{Mor}(C(X))$, $M(u) \in \operatorname{Ob}(C(X))$ denotes the mapping cone of u. $K(X)$ is the category of homotopy equivalence classes of objects of $C(X)$. A diagram in $C(X)$ will be called a *predistinguished triangle,* if its image in $K(X)$ is a distinguished triangle. $D(X)$ denotes the derived category of complexes of $\mathcal{O}_X$-modules. The superscripts $+, -, b$ carry the usual meaning (bounded below, bounded above and bounded). Regarding these notions the basic reference will be [**Hartshorne66**]. S_k denotes the symmetric group of degree k.

Acknowledgements. The more complicated commutative diagrams were drawn with the help of the XY-pic package.

§1. Tools

§§1.1 Wedge products. Let Φ_X and Ψ_X be locally free sheaves on X of rank k and n respectively, and let $\theta_X : \Phi_X \to \Psi_X$ be a morphism.

1.1.1 DEFINITION. Let η be a section of $\bigwedge^p \Psi_X$ over an open set and $\xi_1, \ldots, \xi_k$ sections of Φ_X over the same set. Then $\eta \otimes (\xi_1 \wedge \cdots \wedge \xi_k)$ is a section of $\bigwedge^p \Psi_X \otimes \det \Phi_X$. For any $\sigma \in S_k$ let

$$\xi_{\sigma,q} = \theta_X(\xi_{\sigma(1)}) \wedge \cdots \wedge \theta_X(\xi_{\sigma(q)}),$$

and

$$\xi^{\sigma,q} = \xi_{\sigma(q+1)} \wedge \cdots \wedge \xi_{\sigma(k)}.$$

Further let

$$S_{k,q} = \{\sigma \in S_k | \sigma(1) < \cdots < \sigma(q) \text{ and } \sigma(q+1) < \cdots < \sigma(k)\},$$

and

$$I_\sigma = \{\sigma(1), \ldots, \sigma(q)\}.$$

It is easy to see that every $\sigma \in S_{k,q}$ is determined by I_σ. Now define

$$\lambda_q^\theta(\eta \otimes (\xi_1 \wedge \cdots \wedge \xi_k)) \in \bigwedge^{p+q} \Psi_X \otimes \bigwedge^{k-q} \Phi_X$$

by the formula

$$\lambda_q^\theta(\eta \otimes (\xi_1 \wedge \cdots \wedge \xi_k)) = \sum_{\sigma \in S_{k,q}} (-1)^{\operatorname{sgn}\sigma} (\xi_{\sigma,q} \wedge \eta) \otimes \xi^{\sigma,q},$$

and extend it linearly.

To see that

$$\lambda_q^\theta : \bigwedge^p \Psi_X \otimes \det \Phi_X \to \bigwedge^{p+q} \Psi_X \otimes \bigwedge^{k-q} \Phi_X$$

is a well-defined morphism of sheaves, it is enough to verify the multi-linear and alternating properties. This is left to the reader.

1.1.2 LEMMA. *Let* id *denote* $\mathrm{id}_{\Phi_X} : \Phi_X \to \Phi_X$. *Then*

$$\begin{array}{ccc}
\bigwedge^p \Psi_X \otimes \det \Phi_X \otimes \det \Phi_X & \xrightarrow{\lambda_q^\theta} & \bigwedge^{p+q} \Psi_X \otimes \det \Phi_X \otimes \bigwedge^{k-q} \Phi_X \\
\lambda_{q+r}^\theta \downarrow & & \downarrow \lambda_r^\theta \\
\bigwedge^{p+q+r} \Psi_X \otimes \bigwedge^{k-q-r} \Phi_X \otimes \det \Phi_X & \xrightarrow[\lambda_q^{\mathrm{id}}]{} & \bigwedge^{p+q+r} \Psi_X \otimes \bigwedge^{k-r} \Phi_X \otimes \bigwedge^{k-q} \Phi_X
\end{array}$$

is a commutative diagram, i.e., $\lambda_r^\theta \circ \lambda_q^\theta = \lambda_q^{\mathrm{id}} \circ \lambda_{q+r}^\theta$.

Proof. Use the same notation as in (1.1.1). Then

$$\lambda_r^\theta \circ \lambda_q^\theta \, (\eta \otimes (\xi_1 \wedge \cdots \wedge \xi_k) \otimes (\xi_1 \wedge \cdots \wedge \xi_k)) =$$
$$= \sum_{\tau \in S_{k,r}} \sum_{\sigma \in S_{k,q}} (-1)^{\operatorname{sgn}\tau + \operatorname{sgn}\sigma} (\xi_{\tau,r} \wedge \xi_{\sigma,q} \wedge \eta) \otimes \xi^{\sigma,q} \otimes \xi^{\tau,r}.$$

Let $\sigma \in S_{k,q}$, $\tau \in S_{k,r}$. If

$$I_\tau \cap I_\sigma = \{\tau(1), \ldots, \tau(r)\} \cap \{\sigma(1), \ldots, \sigma(q)\} \neq \emptyset,$$

then $\xi_{\tau,r} \wedge \xi_{\sigma,q} = 0$. Otherwise let $\mu = \mu(\sigma,\tau) \in S_{k,q+r}$ be defined by $I_\mu = I_\tau \cup I_\sigma$ and let $\nu = \nu(\sigma,\tau) = \sigma \in S_{k,q}$.

$$\lambda_q^{\mathrm{id}} \circ \lambda_{q+r}^{\theta} (\eta \otimes (\xi_1 \wedge \cdots \wedge \xi_k) \otimes (\xi_1 \wedge \cdots \wedge \xi_k)) = \sum_{\nu \in S_{k,q}} \sum_{\mu \in S_{k,q+r}} (-1)^{\operatorname{sgn}\mu + \operatorname{sgn}\nu} (\xi_{\mu,q+r} \wedge \eta) \otimes \xi^{\nu,q} \otimes (\xi_{\nu,q} \wedge \xi^{\mu,q+r})$$

and for $\nu \in S_{k,q}$, $\mu \in S_{k,q+r}$, $\xi_{\nu,q} \wedge \xi^{\mu,q+r} \neq 0$ let $\sigma = \sigma(\mu,\nu) = \nu$ and $\tau = \tau(\mu,\nu) \in S_{k,r}$ be defined by $I_\tau = I_\mu \setminus I_\nu$.

This gives a one-to-one correspondence between the pairs (σ,τ) and the pairs (μ,ν).

Observe that

$$(-1)^{\operatorname{sgn}\tau} \underbrace{(\xi_{\tau,r} \wedge \xi_{\sigma,q})}_{\pm\xi_{\mu,q+r}} \otimes \xi^{\tau,r} = (-1)^{\operatorname{sgn}\mu} \xi_{\mu,q+r} \otimes \underbrace{(\xi_{\nu,q} \wedge \xi^{\mu,q+r})}_{\pm\xi^{\tau,r}},$$

so

$$(-1)^{\operatorname{sgn}\tau + \operatorname{sgn}\sigma} (\xi_{\tau,r} \wedge \xi_{\sigma,q} \wedge \eta) \otimes \xi^{\sigma,q} \otimes \xi^{\tau,r} = (-1)^{\operatorname{sgn}\mu + \operatorname{sgn}\nu} (\xi_{\mu,q+r} \wedge \eta) \otimes \xi^{\nu,q} \otimes (\xi_{\nu,q} \wedge \xi^{\mu,q+r}).$$

□

§§**1.2 Hyperfiltrations and spectral sequences.** Let $\mathfrak{A}$ be an abelian category and $D(\mathfrak{A})$ its derived category. Let $\Phi : \mathfrak{A} \to \mathfrak{Ab}$ be a left exact additive functor from $\mathfrak{A}$ to the category of abelian groups and assume that $R\Phi : D(\mathfrak{A}) \to D(\mathfrak{Ab})$, the right derived functor of Φ exists.

1.2.1 DEFINITION. Let $K \in \operatorname{Ob}(D^b(\mathfrak{A}))$ be a bounded complex. A *bounded hyperfiltration* $\mathbb{F}$ of K consists of a set of objects $\mathbb{F}^j K \in \operatorname{Ob}(D^b(\mathfrak{A}))$ for $j = l, \dots, k+1$, where $l, k \in \mathbb{Z}$ and morphisms

$$\varphi_j \in \operatorname{Hom}_{D^b(\mathfrak{A})}(\mathbb{F}^{j+1}K, \mathbb{F}^j K) \qquad \text{for } j = l, \dots, k,$$

where $\mathbb{F}^l K \simeq K$ and $\mathbb{F}^{k+1} K \simeq 0$. $\mathbb{F}^j K$ will be denoted by $\mathbb{F}^j$ when no confusion is likely. For convenience let $\mathbb{F}^i K = K$ for $i < l$ and $\mathbb{F}^i K = 0$ for $i > k$.

The *p-th associated graded complex* of a hyperfiltration $\mathbb{F}$ is

$$\mathbb{G}^p = \mathbb{G}r^p_{\mathbb{F}} K = M(\varphi_p)$$

the mapping cone[1] of the morphism φ_p.

Then one has the following standard result:

[1]Strictly speaking this is the class of the mapping cone of a morphism in $C(X)$ whose class in $D(X)$ is φ_p. The point is that $\mathbb{F}^{p+1} \longrightarrow \mathbb{F}^p \longrightarrow \mathbb{G}^p \xrightarrow{+1}$ forms a distinguished triangle.

1.2.2 THEOREM. *There exists a spectral sequence* E_r *with* $E_1^{p,q} = R^{p+q}\Phi(\mathbb{G}^p)$ *abutting to* $R^{p+q}\Phi(K)$.

The proof of this theorem is relatively straightforward, but cumbersome and is not used elsewhere in this article. The key observation to make in order to prove this result is that one needs to start at the E_1-level since there is no E_0-level, so the proof has to be adjusted accordingly. An important point to notice is the following: once one passes to the E_1-level of this spectral sequence, one leaves the realm of derived categories behind and starts working with long exact sequences of abelian groups (or more generally of objects of an abelian category). This allows one to mimic the usual proof of the similar well-known result for filtrations. For the interested reader the proof is included in an appendix at the end of the article.

1.2.2.1 REMARK. All of the above makes sense and remains true if one replaces $\mathfrak{Ab}$ by an arbitrary abelian category.

§2. Filtration diagrams

Let Φ_X and Ψ_X be locally free sheaves on X of rank k and n respectively, and let $\theta_X : \Phi_X \to \Psi_X$ be a morphism.

Let $p, i \in \mathbb{N}$. We are going to define an object, $\mathfrak{F}_i^p = \mathfrak{F}_i^p(\theta_X) \in \mathrm{Ob}(C(X))$ and a (p,i)*-filtration diagram of* θ_X diagram, $\boxed{\mathfrak{F}_i^p} = \boxed{\mathfrak{F}_i^p(\theta_X)}$. This will be done recursively, starting with $i = 0$ and then increasing i.

2.1 DEFINITION. The $(p,0)$*-filtration diagram of* θ_X is

$$\boxed{\mathfrak{F}_0^p} = \mathfrak{F}_0^p = \bigwedge^{n-p} \Psi_X.$$

A 0*-filtration morphism* for some p, q, consists of locally free sheaves $\mathcal{E}, \mathcal{F}$ and a morphism between $\bigwedge^{n-p} \Psi_X \otimes \mathcal{E}$ and $\bigwedge^{n-q} \Psi_X \otimes \mathcal{F}$.

For instance,

$$\lambda_p^\theta : \bigwedge^{n-p} \Psi_X \otimes \det \Phi_X \to \bigwedge^{n} \Psi_X \otimes \bigwedge^{k-p} \Phi_X$$

is a 0-filtration morphism. Let

$$\mathfrak{F}_1^p = M(\lambda_p^\theta)[-1].$$

2.2 DEFINITION. The $(p,1)$*-filtration diagram of* θ_X consists of the predistinguished triangle,

$$\mathfrak{F}_1^p \to \bigwedge^{n-p} \Psi_X \otimes \det \Phi_X \to \bigwedge^{n} \Psi_X \otimes \bigwedge^{k-p} \Phi_X \xrightarrow{+1}$$

It is denoted by $\boxed{\mathfrak{F}_1^p}$. A 1*-filtration morphism* for some p, r, consists of locally free sheaves $\mathcal{E}, \mathcal{F}$ and morphisms between the corresponding terms of $\boxed{\mathfrak{F}_1^p} \otimes \mathcal{E}$ and

$\boxed{\mathfrak{F}_1^r} \otimes \mathcal{F}$ such that the resulting diagram,

$$\begin{array}{ccc}
\mathfrak{F}_1^p \otimes \mathcal{E} & \longrightarrow & \mathfrak{F}_1^r \otimes \mathcal{F} \\
\downarrow & & \downarrow \\
\bigwedge^{n-p} \Psi_X \otimes \det \Phi_X \otimes \mathcal{E} & \longrightarrow & \bigwedge^{n-r} \Psi_X \otimes \det \Phi_X \otimes \mathcal{F} \\
{\scriptstyle \curlywedge_p^\theta}\downarrow & & \downarrow{\scriptstyle \curlywedge_r^\theta} \\
\bigwedge^{n} \Psi_X \otimes \bigwedge^{k-p} \Phi_X \otimes \mathcal{E} & \longrightarrow & \bigwedge^{n} \Psi_X \otimes \bigwedge^{k-r} \Phi_X \otimes \mathcal{F} \\
\downarrow{\scriptstyle +1} & & \downarrow{\scriptstyle +1}
\end{array}$$

is commutative.

Consider the following commutative diagram (cf. (1.1.2)).

$$\begin{array}{ccc}
\mathfrak{F}_1^p \otimes \det \Phi_X & & \mathfrak{F}_1^{p-q} \otimes \bigwedge^{k-q} \Phi_X \\
\downarrow & & \downarrow \\
\bigwedge^{n-p} \Psi_X \otimes \det \Phi_X \otimes \det \Phi_X & \xrightarrow{\curlywedge_q^\theta} & \bigwedge^{n-p+q} \Psi_X \otimes \det \Phi_X \otimes \bigwedge^{k-q} \Phi_X \\
{\scriptstyle \curlywedge_p^\theta}\downarrow & & \downarrow{\scriptstyle \curlywedge_{p-q}^\theta} \\
\bigwedge^{n} \Psi_X \otimes \bigwedge^{k-p} \Phi_X \otimes \det \Phi_X & \xrightarrow[\curlywedge_q^{\mathrm{id}}]{} & \bigwedge^{n} \Psi_X \otimes \bigwedge^{k-p+q} \Phi_X \otimes \bigwedge^{k-q} \Phi_X \\
\downarrow{\scriptstyle +1} & & \downarrow{\scriptstyle +1}
\end{array} \tag{2.2.1}$$

There exists a morphism,

$$\alpha : \mathfrak{F}_1^p \otimes \det \Phi_X \to \mathfrak{F}_1^{p-q} \otimes \bigwedge^{k-q} \Phi_X,$$

that makes the above diagram commutative.

The diagram (2.2.1), combined with α gives a 1-filtration morphism

$$\boxed{\mathfrak{F}_1^p} \otimes \det \Phi_X \to \boxed{\mathfrak{F}_1^{p-q}} \otimes \bigwedge^{k-q} \Phi_X,$$

with $r = p - q, \mathcal{E} = \det \Phi_X, \mathcal{F} = \bigwedge^{k-q} \Phi_X$.

Let

$$\mathfrak{F}_2^p = M\left(\mathfrak{F}_1^p \otimes \det \Phi_X \to \mathfrak{F}_1^1 \otimes \bigwedge^{k-p+1} \Phi_X \right)[-1].$$

Then there exists a distinguished triangle,

$$\mathfrak{F}_2^p \longrightarrow \mathfrak{F}_1^p \otimes \det \Phi_X \longrightarrow \mathfrak{F}_1^1 \otimes \bigwedge^{k-p+1} \Phi_X \xrightarrow{+1}$$

2.3 Definition. The *$(p,2)$-filtration diagram of* θ_X consists of the diagram,

$$\mathfrak{F}_2^p \longrightarrow \boxed{\mathfrak{F}_1^p} \otimes \det \Phi_X \longrightarrow \boxed{\mathfrak{F}_1^1} \otimes \bigwedge^{k-p+1} \Phi_X.$$

It is denoted by $\boxed{\mathfrak{F}_2^p}$. A *2-filtration morphism* for some p, r, consists of locally free sheaves $\mathcal{E}, \mathcal{F}$ and morphisms between the corresponding terms of $\boxed{\mathfrak{F}_2^p} \otimes \mathcal{E}$ and $\boxed{\mathfrak{F}_2^r} \otimes \mathcal{F}$ such that the resulting diagram is commutative.

More explicitly, the $(p,2)$-filtration diagram of θ_X is:

$$\begin{array}{ccccc}
\mathfrak{F}_2^p & \longrightarrow & \mathfrak{F}_1^p \otimes \det \Phi_X & \longrightarrow & \mathfrak{F}_1^1 \otimes \bigwedge^{k-p+1} \Phi_X \\
 & & \downarrow & & \downarrow \\
 & & \bigwedge^{n-p} \Psi_X \otimes \det \Phi_X \otimes \det \Phi_X & \xrightarrow{\text{⅄}_{p-1}^{\theta}} & \bigwedge^{n-1} \Psi_X \otimes \det \Phi_X \otimes \bigwedge^{k-p+1} \Phi_X \\
 & & \text{⅄}_p^{\theta} \downarrow & & \downarrow \text{⅄}_1^{\theta} \\
 & & \bigwedge^{n} \Psi_X \otimes \bigwedge^{k-p} \Phi_X \otimes \det \Phi_X & \xrightarrow[\text{⅄}_{p-1}^{\mathrm{id}}]{} & \bigwedge^{n} \Psi_X \otimes \bigwedge^{k-1} \Phi_X \otimes \bigwedge^{k-p+1} \Phi_X \\
 & & \downarrow +1 & & \downarrow +1
\end{array}$$

Similarly, a 2-filtration morphism is:

$$
\begin{array}{ccccccccc}
\boxed{\mathfrak{F}_2^p}^{0,0}\otimes\mathcal{E} & \longrightarrow & & & \boxed{\mathfrak{F}_2^p}^{0,1}\otimes\mathcal{E} & \longrightarrow & & & \boxed{\mathfrak{F}_2^p}^{0,2}\otimes\mathcal{E} \\
& \searrow & & & \downarrow & \searrow & & & \downarrow \quad \searrow \\
& & \boxed{\mathfrak{F}_2^r}^{0,0}\otimes\mathcal{F} & \longrightarrow & & & \boxed{\mathfrak{F}_2^r}^{0,1}\otimes\mathcal{F} & \longrightarrow & \boxed{\mathfrak{F}_2^r}^{0,2}\otimes\mathcal{F} \\
& & & & \boxed{\mathfrak{F}_2^p}^{1,1}\otimes\mathcal{E} & \longrightarrow & \downarrow & & \boxed{\mathfrak{F}_2^p}^{1,2}\otimes\mathcal{E} \quad \downarrow \\
& & & & \downarrow & \searrow & & & \downarrow \quad \searrow \\
& & & & & & \boxed{\mathfrak{F}_2^r}^{1,1}\otimes\mathcal{F} & \longrightarrow & \boxed{\mathfrak{F}_2^r}^{1,2}\otimes\mathcal{F} \\
& & & & \boxed{\mathfrak{F}_2^p}^{2,1}\otimes\mathcal{E} & \longrightarrow & \downarrow & & \boxed{\mathfrak{F}_2^p}^{2,2}\otimes\mathcal{E} \quad \downarrow \\
& & & & & \searrow & & & \searrow \\
& & & & & & \boxed{\mathfrak{F}_2^r}^{2,1}\otimes\mathcal{F} & \longrightarrow & \boxed{\mathfrak{F}_2^r}^{2,2}\otimes\mathcal{F},
\end{array}
$$

where the $(p,2)$-filtration diagram,

$$
\begin{array}{ccccc}
\mathfrak{F}_2^p & \longrightarrow & \mathfrak{F}_1^p\otimes\det\Phi_X & \longrightarrow & \mathfrak{F}_1^1\otimes\bigwedge^{k-p+1}\Phi_X \\
& & \downarrow & & \downarrow \\
& & \bigwedge^{n-p}\Psi_X\otimes\det\Phi_X\otimes\det\Phi_X & \longrightarrow & \bigwedge^{n-1}\Psi_X\otimes\det\Phi_X\otimes\bigwedge^{k-p+1}\Phi_X \\
& & \downarrow & & \downarrow \\
& & \bigwedge^{n}\Psi_X\otimes\bigwedge^{k-p}\Phi_X\otimes\det\Phi_X & \longrightarrow & \bigwedge^{n}\Psi_X\otimes\bigwedge^{k-1}\Phi_X\otimes\bigwedge^{k-p+1}\Phi_X \\
& & \downarrow{\scriptstyle +1} & & \downarrow{\scriptstyle +1}
\end{array}
$$

is represented by the simplified diagram,

$$\begin{array}{ccccc}
\boxed{\mathfrak{F}_2^p}^{\,0,0} & \longrightarrow & \boxed{\mathfrak{F}_2^p}^{\,0,1} & \longrightarrow & \boxed{\mathfrak{F}_2^p}^{\,0,2} \\
 & & \big\downarrow & & \big\downarrow \\
 & & \boxed{\mathfrak{F}_2^p}^{\,1,1} & \longrightarrow & \boxed{\mathfrak{F}_2^p}^{\,1,2} \\
 & & \big\downarrow & & \big\downarrow \\
 & & \boxed{\mathfrak{F}_2^p}^{\,2,1} & \longrightarrow & \boxed{\mathfrak{F}_2^p}^{\,2,2}.
\end{array}$$

To define the (p,i)-filtration diagram of θ_X and the i-filtration morphisms we will iterate this construction.

2.4 INDUCTIVE HYPOTHESES For a given i assume that the following holds for all $p, q, r \in \mathbb{N}$.

(2.4.1) The (p,i)*-filtration diagram of* θ_X is defined and denoted by $\boxed{\mathfrak{F}_i^p}$.

(2.4.2) An i*-filtration morphism*, by definition, consists of locally free sheaves $\mathcal{E}, \mathcal{F}$ and a morphism between the corresponding terms of $\boxed{\mathfrak{F}_i^p} \otimes \mathcal{E}$ and $\boxed{\mathfrak{F}_i^r} \otimes \mathcal{F}$ such that the resulting diagram is commutative.

(2.4.3) $\boxed{\mathfrak{F}_i^p}$ has a unique object, $\mathfrak{F}_i^p$, with only one adjacent arrow pointing out.

(2.4.4) $\mathfrak{F}_i^p = 0$ for $p < i$.

(2.4.5) There exists an i-filtration morphism,

$$\curlywedge_q^{\theta,i} : \boxed{\mathfrak{F}_i^p} \otimes \det \Phi_X \to \boxed{\mathfrak{F}_i^{p-q}} \otimes \bigwedge^{k-q} \Phi_X.$$

(2.4.6) The diagram,

$$\begin{array}{ccc}
\boxed{\mathfrak{F}_i^p} \otimes \det \Phi_X \otimes \det \Phi_X & \xrightarrow{\curlywedge_q^{\theta,i}} & \boxed{\mathfrak{F}_i^{p-q}} \otimes \det \Phi_X \otimes \bigwedge^{k-q} \Phi_X \\
{\scriptstyle \curlywedge_{q+r}^{\theta,i}} \big\downarrow & & \big\downarrow {\scriptstyle \curlywedge_r^{\theta,i}} \\
\boxed{\mathfrak{F}_i^{p-q-r}} \otimes \bigwedge^{k-q-r} \Phi_X \otimes \det \Phi_X & \xrightarrow[\curlywedge_q^{\mathrm{id}}]{} & \boxed{\mathfrak{F}_i^{p-q-r}} \otimes \bigwedge^{k-r} \Phi_X \otimes \bigwedge^{k-q} \Phi_X
\end{array}$$

is commutative.

2.5 Lemma-Definition. *If* (2.4) *holds for* $i = 0, \dots, j$, *then* $\boxed{\mathfrak{F}^p_{j+1}}$ *can be defined so that* (2.4) *holds for* $i = j+1$.

Proof. If $j > p$, let $\boxed{\mathfrak{F}^p_{j+1}} = \mathfrak{F}^p_{j+1} = 0$. If $j \leq p$ then by (2.4.5) there exists a j-filtration morphism,

$$\curlywedge^{\theta,j}_{p-j} : \boxed{\mathfrak{F}^p_j} \otimes \det \Phi_X \longrightarrow \boxed{\mathfrak{F}^j_j} \otimes \bigwedge^{k-p+j} \Phi_X.$$

Let $\curlywedge^{\theta,j}_{p-j,res} : \mathfrak{F}^p_j \otimes \det \Phi_X \longrightarrow \mathfrak{F}^j_j \otimes \bigwedge^{k-p+j} \Phi_X$ be the restriction of $\curlywedge^{\theta,j}_{p-j}$ to $\mathfrak{F}^p_j \otimes \det \Phi_X$ and let

$$\mathfrak{F}^p_{j+1} = M(\curlywedge^{\theta,j}_{p-j,res})[-1].$$

Now $\mathfrak{F}^p_{j+1}$ maps to $\mathfrak{F}^p_j$ and they form the predistinguished triangle

$$\mathfrak{F}^p_{j+1} \longrightarrow \mathfrak{F}^p_j \otimes \det \Phi_X \longrightarrow \mathfrak{F}^j_j \otimes \bigwedge^{k-p+j} \Phi_X \xrightarrow{+1} \tag{2.5.1}$$

such that this predistinguished triangle together with $\curlywedge^{\theta,j}_{p-j}$ form a diagram that is commutative.

2.5.2 The $(p, j+1)$-*filtration diagram of* θ_X consists of the diagram

$$\mathfrak{F}^p_{j+1} \longrightarrow \boxed{\mathfrak{F}^p_j} \otimes \det \Phi_X \longrightarrow \boxed{\mathfrak{F}^j_j} \otimes \bigwedge^{k-p+j} \Phi_X.$$

It is denoted by $\boxed{\mathfrak{F}^p_{j+1}}$. A $j+1$-*filtration morphism* for some p, r, consists of locally free sheaves $\mathcal{E}, \mathcal{F}$ and a morphism between the corresponding terms of $\boxed{\mathfrak{F}^p_{j+1}} \otimes \mathcal{E}$ and $\boxed{\mathfrak{F}^r_{j+1}} \otimes \mathcal{F}$ such that the resulting diagram is commutative.

(2.4.1) and (2.4.2) follow from this definition, and (2.4.3) and (2.4.4) are also clear from the construction above.

Next consider the following commutative diagram (cf. (2.4.6)).

$$\begin{array}{ccc}
\mathfrak{F}^{p}_{j+1} \otimes \det \Phi_X & & \mathfrak{F}^{p-q}_{j+1} \otimes \bigwedge^{k-q} \Phi_X \\
\downarrow & & \downarrow \\
\boxed{\mathfrak{F}^{p}_{j}} \otimes \det \Phi_X \otimes \det \Phi_X & \xrightarrow{\curlywedge^{\theta,j}_{q}} & \boxed{\mathfrak{F}^{p-q}_{j}} \otimes \det \Phi_X \otimes \bigwedge^{k-q} \Phi_X \\
\curlywedge^{\theta,j}_{p-j} \downarrow & & \downarrow \curlywedge^{\theta,j}_{p-q-j} \\
\boxed{\mathfrak{F}^{j}_{j}} \otimes \bigwedge^{k-p+j} \Phi_X \otimes \det \Phi_X & \xrightarrow[\curlywedge^{\mathrm{id}}_{q}]{} & \boxed{\mathfrak{F}^{j}_{j}} \otimes \bigwedge^{k-p+q+j} \Phi_X \otimes \bigwedge^{k-q} \Phi_X
\end{array}$$

Then the construction of $\boxed{\mathfrak{F}^{p}_{j+1}}$ implies that there is a morphism,

$$\mathfrak{F}^{p}_{j+1} \otimes \det \Phi_X \longrightarrow \mathfrak{F}^{p-q}_{j+1} \otimes \bigwedge^{k-q} \Phi_X$$

making the diagram commutative and so providing a $j+1$-filtration morphism,

$$\curlywedge^{\theta,j+1}_{q} : \boxed{\mathfrak{F}^{p}_{j+1}} \otimes \det \Phi_X \longrightarrow \boxed{\mathfrak{F}^{p-q}_{j+1}} \otimes \bigwedge^{k-q} \Phi_X. \tag{2.5.3}$$

This proves (2.4.5). Next consider (2.4.6) for $i = j+1$.

$$\begin{array}{ccc}
\boxed{\mathfrak{F}^{p}_{j+1}} \otimes \det \Phi_X \otimes \det \Phi_X & \xrightarrow{\curlywedge^{\theta,j+1}_{q}} & \boxed{\mathfrak{F}^{p-q}_{j+1}} \otimes \bigwedge^{k-q} \Phi_X \otimes \det \Phi_X \\
\curlywedge^{\theta,j+1}_{q+r} \downarrow & & \downarrow \curlywedge^{\theta,j+1}_{r} \\
\boxed{\mathfrak{F}^{p-q-r}_{j+1}} \otimes \det \Phi_X \otimes \bigwedge^{k-q-r} \Phi_X & \xrightarrow[\curlywedge^{\mathrm{id}}_{q}]{} & \boxed{\mathfrak{F}^{p-q-r}_{j+1}} \otimes \bigwedge^{k-q} \Phi_X \otimes \bigwedge^{k-r} \Phi_X
\end{array} \tag{2.5.4}$$

This is shown in more detail in the following diagram. Due to the size of the diagram, it is broken up into two pieces.

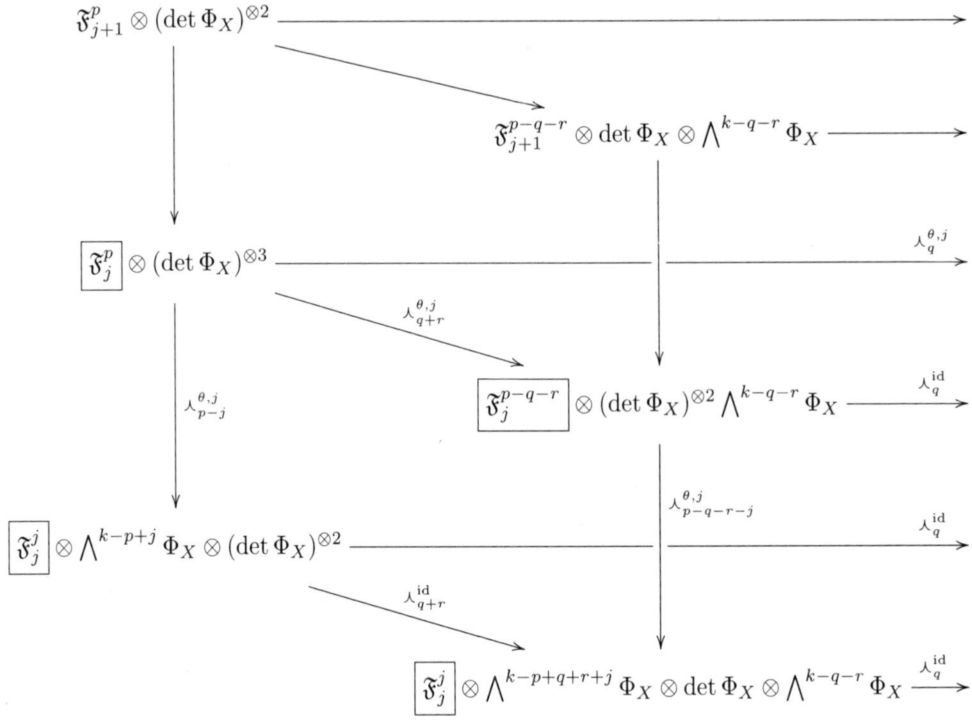
$\mathfrak{F}^{p}_{j+1} \otimes (\det \Phi_X)^{\otimes 2}$
$\mathfrak{F}^{p-q-r}_{j+1} \otimes \det \Phi_X \otimes \bigwedge^{k-q-r} \Phi_X$
$\boxed{\mathfrak{F}^{p}_{j}} \otimes (\det \Phi_X)^{\otimes 3}$
$\lambda^{\theta,j}_{q}$
$\lambda^{\theta,j}_{q+r}$
$\lambda^{\theta,j}_{p-j}$
$\boxed{\mathfrak{F}^{p-q-r}_{j}} \otimes (\det \Phi_X)^{\otimes 2} \bigwedge^{k-q-r} \Phi_X$
$\lambda^{\mathrm{id}}_{q}$
$\lambda^{\theta,j}_{p-q-r-j}$
$\boxed{\mathfrak{F}^{j}_{j}} \otimes \bigwedge^{k-p+j} \Phi_X \otimes (\det \Phi_X)^{\otimes 2}$
$\lambda^{\mathrm{id}}_{q}$
$\lambda^{\mathrm{id}}_{q+r}$
$\boxed{\mathfrak{F}^{j}_{j}} \otimes \bigwedge^{k-p+q+r+j} \Phi_X \otimes \det \Phi_X \otimes \bigwedge^{k-q-r} \Phi_X$
$\lambda^{\mathrm{id}}_{q}$

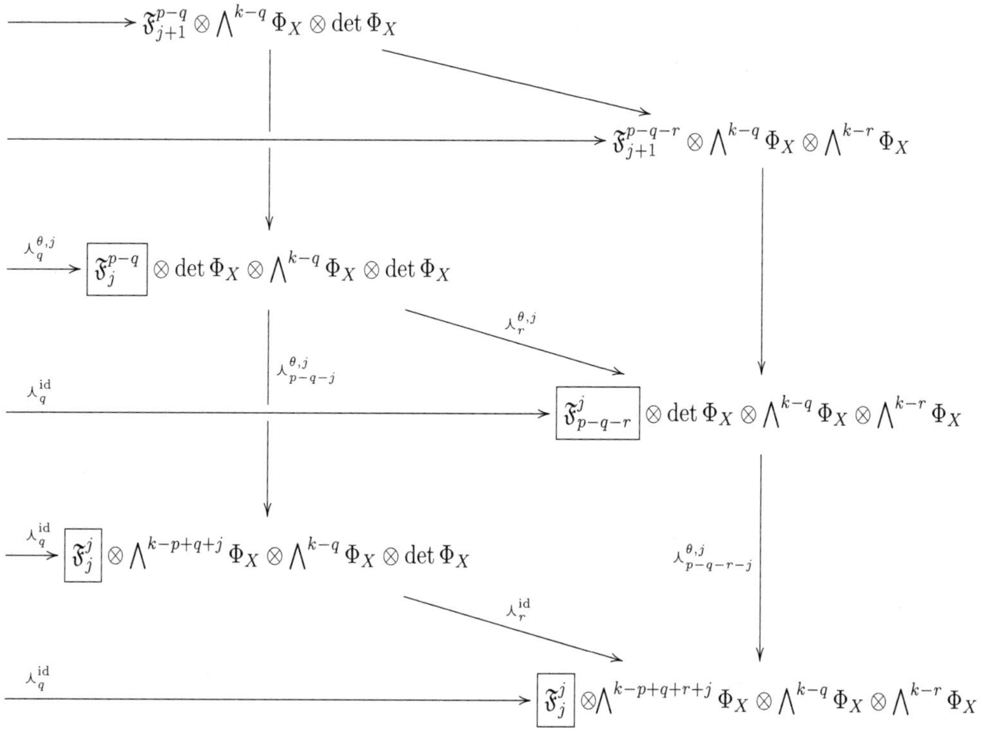

The commutativity of vertical faces is either obvious or follows by the commutativity of (2.5.3).

$$\begin{array}{ccc}
\boxed{\mathfrak{F}_j^p} \otimes (\det \Phi_X)^{\otimes 3} & \xrightarrow{\curlywedge_q^{\theta,j}} & \boxed{\mathfrak{F}_j^{p-q}} \otimes \det \Phi_X \otimes \bigwedge^{k-q} \Phi_X \otimes \det \Phi_X \\
\Big\downarrow {\scriptstyle \curlywedge_{q+r}^{\theta,j}} & & \Big\downarrow {\scriptstyle \curlywedge_r^{\theta,j}} \\
\boxed{\mathfrak{F}_j^{p-q-r}} \otimes (\det \Phi_X)^{\otimes 2} \otimes \bigwedge^{k-q-r} \Phi_X & \xrightarrow[\curlywedge_q^{\mathrm{id}}]{} & \boxed{\mathfrak{F}_j^{p-q-r}} \otimes \det \Phi_X \otimes \bigwedge^{k-q} \Phi_X \otimes \bigwedge^{k-r} \Phi_X
\end{array}$$

is commutative by (2.4.6), and

$$\begin{array}{ccc}
\boxed{\mathfrak{F}^j_j}\otimes \bigwedge^{k-p+j}\Phi_X\otimes(\det\Phi_X)^{\otimes 2} & \xrightarrow{\lambda^{\mathrm{id}}_q} & \boxed{\mathfrak{F}^j_j}\otimes \bigwedge^{k-p+q+j}\Phi_X\otimes \bigwedge^{k-q}\Phi_X\otimes\det\Phi_X \\
\lambda^{\mathrm{id}}_{q+r}\Big\downarrow & & \Big\downarrow\lambda^{\mathrm{id}}_r \\
\boxed{\mathfrak{F}^j_j}\otimes \bigwedge^{k-p+q+r+j}\Phi_X\otimes\det\Phi_X\otimes \bigwedge^{k-q-r}\Phi_X & \xrightarrow[\lambda^{\mathrm{id}}_q]{} & \boxed{\mathfrak{F}^j_j}\otimes \bigwedge^{k-p+q+r+j}\Phi_X\otimes \bigwedge^{k-q}\Phi_X\otimes \bigwedge^{k-r}\Phi_X
\end{array}$$

is commutative by (1.1.2). Now it is easy to see that (2.5.4) is commutative. □

2.6. We are ready to define $\underline{\mathfrak{Q}}^p_{\theta_X}\in \mathrm{Ob}(D(X))$ for $p\in\mathbb{Z}, p\geq -k$. Let $\underline{\mathfrak{Q}}^p_{\theta_X}$ be the class of

$$\mathfrak{F}^{n-k-p}_{n-k-p}\otimes(\det\Phi_X)^{-(n-k-p+1)}$$

in $\mathrm{Ob}(D(X))$ for $-k\leq p\leq n-k$, and let $\underline{\mathfrak{Q}}^p_{\theta_X}=0$ for $p>n-k$. It is easy to see that

$$\underline{\mathfrak{Q}}^{n-k}_{\theta_X}=\det\Psi_X\otimes(\det\Phi_X)^{-1}$$

and that there is a distinguished triangle:

$$\underline{\mathfrak{Q}}^{n-k-1}_{\theta_X}\otimes\det\Phi_X\to \bigwedge^{n-1}\Psi_X\to \underline{\mathfrak{Q}}^{n-k}_{\theta_X}\otimes\bigwedge^{k-1}\Phi_X\xrightarrow{+1}$$

In general for $j\geq p-n+k$ let $\mathbb{F}^j\bigwedge^p\Psi_X$ be the class of

$$\mathfrak{F}^{n-p}_{n-k-p+j}\otimes(\det\Phi_X)^{-(n-k-p+j)}$$

in $\mathrm{Ob}(D(X))$. The predistinguished triangle (2.5.1),

$$\mathfrak{F}^{n-p}_{n-k-p+j+1}\longrightarrow \mathfrak{F}^{n-p}_{n-k-p+j}\otimes\det\Phi_X\longrightarrow \mathfrak{F}^{n-k-p+j}_{n-k-p+j}\otimes\bigwedge^{j}\Phi_X\xrightarrow{+1}$$

gives the distinguished triangle,

$$\mathbb{F}^{j+1}\bigwedge^p\Psi_X\to\mathbb{F}^{j}\bigwedge^p\Psi_X\to\underline{\mathfrak{Q}}^{p-j}_{\theta_X}\otimes\bigwedge^j\Phi_X\xrightarrow{+1}.$$

Now $\mathbb{F}^{k+1}\bigwedge^p\Psi_X=0$ by (2.4.4) and by definition $\mathbb{F}^{p-n+k}\bigwedge^p\Psi_X=\bigwedge^p\Psi_X$.

Observe that if $p-n+k<0$, then

$$\mathbb{F}^0\bigwedge^p \Psi_X \simeq \mathbb{F}^{-1}\bigwedge^p \Psi_X \simeq \cdots \simeq \mathbb{F}^{p-n+k}\bigwedge^p \Psi_X = \bigwedge^p \Psi_X,$$

since $\bigwedge^j \Phi_X = 0$ for $j<0$. If $p-n+k \geq 0$, define $\mathbb{F}^j\bigwedge^p \Psi_X = \bigwedge^p \Psi_X$ for $j=0,\dots,p-n+k$.

Therefore we proved:

2.7 THEOREM. *Let Φ_X and Ψ_X be locally free sheaves on X of rank k and n respectively, and let $\theta_X : \Phi_X \to \Psi_X$ be a morphism. Then there exists a $\underline{\mathfrak{Q}}^r_{\theta_X} \in$ $\mathrm{Ob}(D(X))$ for all $r\in\mathbb{Z}, r\geq -k$ with the following property. For any $p\in\mathbb{N}$ there exists a hyperfiltration $\mathbb{F}^j\bigwedge^p \Psi_X$ of $\bigwedge^p \Psi_X$ with $j=0,\dots,k+1$, such that*

$$\mathbb{F}^0\bigwedge^p \Psi_X \simeq \bigwedge^p \Psi_X,$$

$$\mathbb{F}^{k+1}\bigwedge^p \Psi_X \simeq 0$$

and

$$\mathbb{G}^j\bigwedge^p \Psi_X \simeq \underline{\mathfrak{Q}}^{p-j}_{\theta_X} \otimes \bigwedge^j \Phi_X.$$

Furthermore, for $r>n-k$,

$$\underline{\mathfrak{Q}}^r_{\theta_X} \simeq 0.$$

Now using (1.2.2) one obtains the following corollary:

2.8 COROLLARY. *Let Φ_X and Ψ_X be locally free sheaves on X of rank k and n respectively, and let $\theta_X : \Phi_X \to \Psi_X$ be a morphism. Then for any locally free sheaf $\mathcal{E}$ and for any $p\in\mathbb{N}$ there exists a spectral sequence,*

$$E_1^{r,s} = \mathbb{H}^{r+s}(X, \underline{\mathfrak{Q}}^{p-r}_{\theta_X}\otimes \bigwedge^r \Phi_X \otimes \mathcal{E}) \Rightarrow H^{r+s}(X, \bigwedge^p \Psi_X \otimes \mathcal{E}).$$

2.9 PROPOSITION. *If the cokernel sheaf, Ξ_X, of θ_X is locally free, then $\underline{\mathfrak{Q}}^p_{\theta_X}$ is simply the p-th exterior power of Ξ_X, and the spectral sequence above is the well-known spectral sequence associated to the filtration,*

$$\bigwedge^p \Psi_X = F^0 \supset F^1 \supset \cdots \supset F^p \supset F^{p+1} = 0,$$

with quotients

$$F^j/F^{j+1} \simeq \bigwedge^{p-j} \Xi_X \otimes \bigwedge^j \Phi_X$$

for each j.

Proof. By definition one has that

$$\underline{\Omega}^p_{\theta_X} \simeq \det \Psi_X \otimes (\det \Phi_X)^{-1} \simeq \det \Xi_X.$$

Then the statement follows using descending induction, the filtration associated to the short exact sequence of locally free sheaves and the distinguished triangle,

$$\mathbb{F}^{j+1} \bigwedge^p \Psi_X \longrightarrow \mathbb{F}^j \bigwedge^p \Psi_X \longrightarrow \underline{\Omega}^{p-j}_{\theta_X} \otimes \bigwedge^j \Phi_X \xrightarrow{+1} .$$

□

2.10 Example. Let $k = 1$, i.e., assume that Φ_X is a line bundle. Then the hyperfiltration in (2.7) is simply a distinguished triangle,

$$\underline{\Omega}^{p-1}_{\theta_X} \otimes \Phi_X \longrightarrow \bigwedge^p \Psi_X \longrightarrow \underline{\Omega}^p_{\theta_X} \xrightarrow{+1} .$$

and then the spectral sequence in (2.8) reduces to a long exact hypercohomology sequence:

$$\dots \longrightarrow \mathbb{H}^s(X, \underline{\Omega}^{p-1}_{\theta_X} \otimes \Phi_X \otimes \mathcal{E}) \longrightarrow H^s(X, \bigwedge^p \Psi_X \otimes \mathcal{E}) \longrightarrow$$

$$\longrightarrow \mathbb{H}^s(X, \underline{\Omega}^p_{\theta_X} \otimes \Phi_X \otimes \mathcal{E}) \longrightarrow \mathbb{H}^{s+1}(X, \underline{\Omega}^{p-1}_{\theta_X} \otimes \Phi_X \otimes \mathcal{E}) \longrightarrow \dots$$

§3. A vanishing theorem

Throughout this section X is an algebraic veriety over a field k.

The following well-known fact is included for ease of reference.

3.1 Fact. *Let $\mathcal{E}$ be a locally free sheaf of rank r. Assume that there is a filtration*

$$\mathcal{E} = F^0 \supset F^1 \supset \dots \supset F^r = 0$$

of $\mathcal{E}$ such that

$$F^{i-1}/F^i = \mathcal{L}_i$$

is a line bundle for all $i = 1, \dots, r$. Then for every $1 \le t \le r$ there is a filtration

$$\bigwedge^t \mathcal{E} = F^0_t \supset F^1_t \supset \dots \supset F^{\binom{r}{t}}_t = 0$$

of $\bigwedge^t \mathcal{E}$ such that

$$F^{i-1}_t / F^i_t = \mathcal{L}_{i_1} \otimes \mathcal{L}_{i_2} \otimes \dots \otimes \mathcal{L}_{i_t}$$

for all $i = 1, \dots, \binom{r}{t}$ and a suitable set of indicies $1 \le i_1 < i_2 < \dots < i_t \le r$. □

3.2 DEFINITION. A locally free sheaf $\mathcal{E}$ on X is called *semi-positive* or *nef* if for every smooth proper curve C and every map $\gamma : C \to X$, $\deg \mathcal{Q}\big|_C \geq 0$ for any quotient bundle $\mathcal{Q}$ of $\gamma^*\mathcal{E}$.

3.3 DEFINITION. Let $\mathcal{E}$ be a locally free sheaf of rank r. $\mathcal{E}$ will be called *semi-negative of splitting type* if $\mathcal{E}$ has a filtration

$$\mathcal{E} = F^0 \supset F^1 \supset \cdots \supset F^r = 0$$

such that

$$F^{i-1}/F^i = \mathcal{L}_i$$

is a semi-negative line bundle, i.e., $\mathcal{L}_i^{-1}$ is nef.

3.4 DEFINITION. A class, $\mathfrak{C}$, of line bundles on X will be called *nef-invariant*, if for all $\mathcal{L} \in \mathfrak{C}$ and $\mathcal{N}$ nef line bundle on X, $\mathcal{L} \otimes \mathcal{N} \in \mathfrak{C}$.

3.5 EXAMPLE. If X is projective, then the class of ample line bundles on X is nef-invariant.

3.6 LEMMA. *Let Φ_X and Ψ_X be locally free sheaves on X of rank k and n respectively, and let $\theta_X : \Phi_X \to \Psi_X$ be a morphism. Let $\mathfrak{C}$ be a nef-invariant class of line bundles on X. Let $c(n) \in \mathbb{N}$ be a constant such that $H^p(X, \bigwedge^q \Psi_X \otimes \mathcal{L}) = 0$ for every line bundle $\mathcal{L} \in \mathfrak{C}$ and $p + q > c(n)$. Let $i, j \in \mathbb{N}$ be natural numbers such that $i + j + k \geq c(n)$. Assume that Φ_X is semi-negative of splitting type and that for every natural number $l > j$ and line bundle $\mathcal{L} \in \mathfrak{C}$ on X,*

$$\mathbb{H}^i(X, \underline{\mathfrak{Q}}^l_{\theta_X} \otimes \det \Phi_X \otimes \mathcal{L}) = 0.$$

Then

$$\mathbb{H}^{i+1}(X, \underline{\mathfrak{Q}}^j_{\theta_X} \otimes \det \Phi_X \otimes \mathcal{L}) = 0$$

for every line bundle $\mathcal{L} \in \mathfrak{C}$.

Proof. By (2.8) there exists a spectral sequence

$$E_1^{r,s} = \mathbb{H}^{r+s}(X, \underline{\mathfrak{Q}}^{j+k-r}_{\theta_X} \otimes \bigwedge^r \Phi_X \otimes \mathcal{L}) \Rightarrow H^{r+s}(X, \bigwedge^{j+k} \Psi_X \otimes \mathcal{L}).$$

Since $i+j+k \geq c(n)$, $H^{i+1}(X, \bigwedge^{j+k} \Psi_X \otimes \mathcal{L}) = 0$ by assumption. Hence $E_\infty^{r,i+1-r} = 0$ for all r. In particular $E_\infty^{k,i+1-k} = 0$. Suppose now that

$$E_1^{k,i+1-k} = \mathbb{H}^{i+1}(X, \underline{\mathfrak{Q}}^j_{\theta_X} \otimes \det \Phi_X \otimes \mathcal{L}) \neq 0.$$

Observe that $E_w^{u,v} = 0$ for every $u > k$ and arbitrary v, w, so in order to have $E_\infty^{k,i+1-k} = 0$, there must be a $t \geq 1$ such that $E_t^{k-t,i-k+t} \neq 0$. Then $E_1^{k-t,i-k+t} \neq 0$ for the same t, so $l = j + t > j$ is such that

$$E_1^{k-t,i-k+t} = \mathbb{H}^i(X, \underline{\mathfrak{Q}}^l_{\theta_X} \otimes \bigwedge^{k-t} \Phi_X \otimes \mathfrak{L}) \neq 0.$$

Now by assumption Φ_X has a filtration

$$\Phi_X = F^0 \supset F^1 \supset \cdots \supset F^k = 0$$

such that $F^{i-1}/F^i = \mathfrak{L}_i$ and $\mathfrak{L}_i^{-1}$ is a nef line bundle. Then by (3.1) there exist $i_1, \ldots, i_{k-t}$ such that

$$\mathbb{H}^i(X, \underline{\mathfrak{Q}}^l_{\theta_X} \otimes \mathfrak{L}_{i_1} \otimes \cdots \otimes \mathfrak{L}_{i_{k-t}} \otimes \mathfrak{L}) \neq 0.$$

Therefore

$$\mathbb{H}^i(X, \underline{\mathfrak{Q}}^l_{\theta_X} \otimes \det \Phi_X \otimes \mathfrak{L} \otimes \mathfrak{L}^{-1}_{i_{k-t+1}} \otimes \cdots \otimes \mathfrak{L}^{-1}_{i_k}) \neq 0,$$

for $\{i_{k-t+1}, \ldots, i_k\} = \{1, \ldots, k\} \setminus \{i_1, \ldots, i_{k-t}\}$.

Since $\mathfrak{L} \otimes \mathfrak{L}^{-1}_{i_{k-t+1}} \otimes \cdots \otimes \mathfrak{L}^{-1}_{i_k} \in \mathfrak{C}$, this non-vanishing violates the assumption. Hence the statement follows. □

3.7 THEOREM. *Let Φ_X and Ψ_X be locally free sheaves on X of rank k and n respectively, and let $\theta_X : \Phi_X \to \Psi_X$ be a morphism. Let $\mathfrak{C}$ be a nef-invariant class of line bundles on X. Assume that $H^p(X, \bigwedge^q \Psi_X \otimes \mathfrak{L}) = 0$ for every line bundle $\mathfrak{L} \in \mathfrak{C}$ and $p + q > n$. Assume further that Φ_X is semi-negative of splitting type. Then for every line bundle $\mathfrak{L} \in \mathfrak{C}$,*

$$\mathbb{H}^i(X, \underline{\mathfrak{Q}}^l_{\theta_X} \otimes \det \Phi_X \otimes \mathfrak{L}) = 0 \qquad \textit{for } i + l > n - k.$$

Proof. By (2.7) $\underline{\mathfrak{Q}}^l_{\theta_X} = 0$ for $l > n - k$, so

$$\mathbb{H}^i(X, \underline{\mathfrak{Q}}^l_{\theta_X} \otimes \det \Phi_X \otimes \mathfrak{L}) = 0$$

for $l > n - k$ and $i \geq 0$. Then

$$\mathbb{H}^{i+1}(X, \underline{\mathfrak{Q}}^{n-k}_{\theta_X} \otimes \det \Phi_X \otimes \mathfrak{L}) = 0$$

for $i \geq 0$ by (3.6). Hence

$$\mathbb{H}^i(X, \underline{\mathfrak{Q}}^l_{\theta_X} \otimes \det \Phi_X \otimes \mathfrak{L}) = 0$$

for $l > n - k - 1$ and $i \geq 1$. Then again

$$\mathbb{H}^{i+1}(X, \underline{\mathfrak{Q}}^{n-k-1}_{\theta_X} \otimes \det \Phi_X \otimes \mathfrak{L}) = 0$$

for $i \geq 1$ by (3.6). Hence again

$$\mathbb{H}^i(X, \underline{\mathfrak{Q}}^l_{\theta_X} \otimes \det \Phi_X \otimes \mathcal{L}) = 0$$

for $l > n - k - 2$ and $i \geq 2$.

Iterating this process one sees that

$$\mathbb{H}^i(X, \underline{\mathfrak{Q}}^l_{\theta_X} \otimes \det \Phi_X \otimes \mathcal{L}) = 0$$

for $i + l > n - k$. □

3.8 EXAMPLE. If X is a smooth projective variety over $\mathbb{C}$ and $\mathfrak{C}$ is the class of ample line bundles on X, then $\Psi_X = \Omega_X$ satisfies the assumption by the Kodaira-Akizuki-Nakano vanishing theorem.

3.9 COROLLARY. **[Kovács97a, 1.4]** *Let $f : X \to S$ be a smooth morphism of smooth projective varieties over $\mathbb{C}$ of dimension n and k respectively. Let $\mathcal{L}$ be an ample line bundle and assume that $f^*\Omega_S$ is semi-negative of splitting type. Then*

$$H^i(X, \Omega^l_{X/S} \otimes f^*\omega_S \otimes \mathcal{L}) = 0 \qquad \textit{for } i + l > n - k.$$

Proof. (3.8), (3.7), (2.9.1). □

§4. Functoriality

Let S be a scheme and $\mathfrak{Sch}_S$ the category of S-schemes. Let $\mathfrak{LocFree}_S$ be the category of pairs $(X, \mathcal{E})$ where $X \in \mathrm{Ob}(\mathfrak{Sch}_S)$ and $\mathcal{E}$ is a locally free sheaf on X. A morphism in $\mathrm{Hom}_{\mathfrak{LocFree}_S}((Y, \mathcal{F}), (X, \mathcal{E}))$ consists of pairs (ϕ, φ) where $\phi \in \mathrm{Hom}_{\mathfrak{Sch}_S}(Y, X)$ and $\varphi \in \mathrm{Hom}_X(\mathcal{E}, \phi_*\mathcal{F}) \simeq \mathrm{Hom}_Y(\phi^*\mathcal{E}, \mathcal{F})$.

A functor $\Phi = (\Phi_1, \Phi_2) : \mathfrak{Sch}_S \to \mathfrak{LocFree}_S$ will be called *honest* if Φ_1 is the identity. In this case Φ_X will denote $\Phi_2(X)$, i.e., Φ_X is a locally free sheaf on X.

4.1 THEOREM. *Let Φ and Ψ be honest functors from $\mathfrak{Sch}_S$ to $\mathfrak{LocFree}_S$, and $\theta : \Phi \to \Psi$ a natural transformation. Then the definition of $\mathbb{F}^q \bigwedge^p \Psi_X$ and $\underline{\mathfrak{Q}}^p_{\theta_X}$ is functorial in the following sense, if $\phi : Y \to X$ is a morphism of S-schemes, such that $\mathrm{rk}\, \Phi_X = \mathrm{rk}\, \Phi_Y = k$, then there are natural maps in $D(X)$ forming the commutative diagram:*

$$\begin{array}{ccccccc}
\mathbb{F}^{q+1}\bigwedge^p \Psi_X & \longrightarrow & \mathbb{F}^q\bigwedge^p \Psi_X & \longrightarrow & \mathbb{G}^q\bigwedge^p \Psi_X & \xrightarrow{+1} & \\
\downarrow & & \downarrow & & \downarrow & & \\
R\phi_*\mathbb{F}^{q+1}\bigwedge^p \Psi_Y & \longrightarrow & R\phi_*\mathbb{F}^q\bigwedge^p \Psi_Y & \longrightarrow & R\phi_*\mathbb{G}^q\bigwedge^p \Psi_Y & \xrightarrow{+1} & .
\end{array}$$

In particular, there are natural maps in $D(X)$ forming the commutative diagram:

$$\begin{array}{ccc} \bigwedge^p \Psi_X & \longrightarrow & \underline{\Omega}^p_{\theta_X} \\ \downarrow & & \downarrow \\ R\phi_* \bigwedge^p \Psi_Y & \longrightarrow & R\phi_* \underline{\Omega}^p_{\theta_Y}. \end{array}$$

Proof. Notice that θ induces a natural transformation, $\theta^p : \bigwedge^p \Phi \to \bigwedge^p \Psi$, between the functors $\bigwedge^p \Phi : X \longmapsto (X, \bigwedge^p \Phi_X)$ and $\bigwedge^p \Psi : X \longmapsto (X, \bigwedge^p \Psi_X)$. Then add the following to (2.4.5):

(2.4.5′) $\lambda_q^{\theta,i}$ is natural, i.e., there exists a commutative diagram:

$$\begin{array}{ccc} \boxed{\mathfrak{F}_i^p(\theta_X)} \otimes \det \Phi_X & \xrightarrow{\lambda_q^{\theta,i}} & \boxed{\mathfrak{F}_i^{p-q}(\theta_X)} \otimes \bigwedge^{k-q} \Phi_X \\ \Big\downarrow {\scriptstyle \theta^k(\phi)} & & \Big\downarrow {\scriptstyle \theta^{k-q}(\phi)} \\ \boxed{\mathfrak{F}_i^p(\theta_Y)} \otimes \det \Phi_Y & \xrightarrow{\lambda_q^{\theta,i}} & \boxed{\mathfrak{F}_i^{p-q}(\theta_Y)} \otimes \bigwedge^{k-q} \Phi_Y. \end{array}$$

Now repeating the proof of (2.5) for X and Y simultaneously it is easy to see that the first statement is a simple consequence of (2.4.5′) and the construction of $\mathbb{F}^q \bigwedge^p \Psi_X$. Then the second statement follows directly from (2.7). □

4.1.1 Remark. In the above statements "natural" should be understood the following way: In the definition of $\underline{\Omega}^p_{\theta_X}$ one needs to make non-canonical choices when defining objects as the third vertex of a distinguished triangle. Here "natural" means that once those choices are made, the construction of the above maps is natural.

§5. Applications

The common theme of the applications is that certain arguments that involve ample line bundles and vanishing theorems on smooth varieties can be extended to certain singular varieties. These results about singular varieties in turn play an important role toward proving other results about smooth varieties and the key line bundles being only nef and big instead of ample.

5.1 Definition. A line bundle $\mathcal{K}$ on Y is called *big* if Y is proper and $\mathcal{K}^{\otimes m}$ gives a birational map for some $m > 0$.

5.2 Example. Let $f : Y \to X$ be a birational morphism between projective varieties and let $\mathcal{L}$ be an ample line bundle on X. Then $f^*\mathcal{L}$ is nef and big on Y.

For the rest of this section we assume that everything is defined over $\mathbb{C}$.

§§**5.3.** Let S be a smooth variety. Let Ψ be the functor of Kähler differentials, i.e., for any S-scheme X, smooth over $\mathbb{C}$, $\Psi_X = \Omega_X$, i.e., $\Psi(X) = (X, \Omega_X)$, and let Φ be the functor on $\mathfrak{Sch}_S$ that is the pullback of Ψ from S, i.e., for $f : X \longrightarrow S$, $\Phi_X = f^*\Omega_S$, or in other words $\Phi(X) = (X, f^*\Omega_S)$. Then f^* induces a natural transformation $\theta : \Phi \longrightarrow \Psi$.

Let $f : X \longrightarrow S$ be a morphism of algebraic varieties of dimension n and k respectively, such that S is smooth. Then one obtains the definition of the p-th relative De Rham complexes[2], $\underline{\Omega}^p_{X/S} = \underline{\mathfrak{Q}}^p_{\theta_X}$, for non-smooth morphisms. The germ of this construction first appeared in [**Kovács96**], but there the base was required to be a smooth curve. The initial motivation to define the relative De Rham complexes was to prove the following statement:

5.3.1 THEOREM. [**Kovács96**] *Let $g : Y \longrightarrow C$ be a smooth family of projective varieties of general type with a nef canonical bundle and C a smooth projective curve of genus at most one. Then the fibers of g are birational.*

5.3.1.1 REMARK. The case of $\dim Y = 2$ is well-known and the case of $\dim Y = 3$ was first proved by [**Migliorini95**].

The construction presented here allows the generalization of the p-th relative De Rham complexes to the case when the base of the family is higher dimensional (but still smooth). This is discussed and the following generalization of (5.3.1) is proved in [**Kovács97b**].

5.3.2 THEOREM. [**Kovács97b**] *Let $g : Y \longrightarrow S$ be a smooth family of projective varieties of general type with a nef canonical bundle and S a smooth projective variety. Assume that Ω_S is semi-negative. Then the fibres of g are birational.*

Further generalizations were obtained in [**Kovács02**], where logarithmic analogues of the above were constructed. For details the reader is referred to the original article.

§6. The 'nef & big on smooth' $\approx$ 'ample on singular' principal

It may not be entirely clear at this point how the results of the present article are applied to prove the above theorems, or why they are necessary at all. The statements are about smooth varieties whereas the results in this paper are only interesting for singular ones.

The main point is that without these results the arguments that are used in the applications would only prove the statements for *ample* line bundles. By working with singular varieties we are able to extend the argument to include *nef and big* line bundles as explained below.

The essence of the applications may be summarized as the

"Nef & big line bundle on smooth $\approx$ ample line bundle on singular" principle.

In fact, it is a new approach to this principle:

---

[2]Note that the word *complex* here refers to an object in a derived category. These are analogues of the *components* of the usual De Rham complex and not the total complex. For more details please refer to [**Kovács97b**].

The usual approach sets its goal to obtain a statement for a possibly singular X and an ample line bundle $\mathcal{L}$ on X. This is achieved by working with a desingularization, $f : Y \to X$, and a nef and big line bundle, $f^*\mathcal{L}$, on Y.

Our "upside down" approach aims for a statement for a smooth Y and a nef and big line bundle $\mathcal{K}$ on Y. If $\mathcal{K}^{\otimes m}$ is generated by global sections, then we may work with the morphism, $f : Y \to X$, induced by $\mathcal{K}^{\otimes m}$, so $\mathcal{K}^{\otimes m} = f^*\mathcal{L}$ for some $\mathcal{L}$ ample on X. The main idea is that working on X we may assume that our line bundle is ample. The price to pay is that we no longer can assume that the variety is smooth, but fortunately this is not a fatal problem.

The detailed plan is the following:

6.1 PLAN

(6.1.1) First prove the statement for Y smooth, $\mathcal{K}$ ample.
(6.1.2) Check if $\mathcal{K}^{\otimes m}$ is generated by global sections. This gives $f : Y \to X$ induced by $\mathcal{K}^{\otimes m}$, so $\mathcal{K}^{\otimes m} = f^*\mathcal{L}$ for some $\mathcal{L}$ ample on X.
(6.1.3) Extend the argument (from (6.1.1)) for X singular, $\mathcal{L}$ ample.
(6.1.4) "Pull back" the result.

Next we will sketch how this plan is executed in the applications mentioned in the previous section. Let us remark that the line bundle $\mathcal{K}$ is chosen to be the canonical bundle, ω_Y.

6.2. The key ingredient for (6.1.1) is the spectral sequence from the introduction. For more details please refer to the actual applications.

6.3. (6.1.2) follows easily from the Basepoint-free Theorem [**CKM88, 9.3**]. It provides the needed global generation for a power of the canonical bundle.

6.4. To complete (6.1.3) the necessary ingredients are the argument from (6.1.1), (2.7) and (2.8). In addition we use that if $g : Y \to S$ is as in (5.3.2), then $\underline{\Omega}^j_{Y/S} \simeq \Omega^j_{Y/S}$ by (2.9).

6.5. Functoriality becomes important for (6.1.4):
For the morphism $f : Y \to X$ induced by $\omega_Y^{\otimes m}$ and for all j there exists a natural morphism,

$$\phi^* : \underline{\Omega}^j_{X/S} \to R\phi_* \underline{\Omega}^j_{Y/S}.$$

One can also prove that there exists a natural morphism $\rho : \mathcal{O}_X \to \underline{\Omega}^0_{X/S}$ cf. [**Kovács97b**].

Next we need the following:

6.6 DEFINITION. [**Kovács97b**] $f : X \to S$ is an *SP-morphism* if

$$\rho : \mathcal{O}_X \to \underline{\Omega}^0_{X/S}$$

has a *left inverse* in $D(X)$, i.e., $\exists$ a morphism $\rho' : \underline{\Omega}^0_{X/S} \to \mathcal{O}_X$ such that $\rho' \circ \rho : \mathcal{O}_X \to \mathcal{O}_X$ is a quasi-isomorphism.

6.7 OBSERVATION. *If $f : X \to S$ is an SP-morphism, then*

$$H^i(X,\mathcal{L}) = \mathbb{H}^i(X, \mathcal{O}_X \otimes \mathcal{L}) \subseteq \mathbb{H}^i(X, \underline{\Omega}^0_{X/S} \otimes \mathcal{L})$$

for all i and for all locally free sheaves $\mathcal{L}$.

6.8 EXAMPLE. If f is smooth it is SP.

The importance of this notion is shown by the next simple lemma.

6.9 LEMMA.

$$\begin{array}{ccc} Y & \xrightarrow{\phi} & X \\ {\scriptstyle g}\downarrow & & \downarrow{\scriptstyle f} \\ S & = & S \end{array}$$

Let S and g be smooth, ϕ birational and assume that X has rational singularities. Then f is an SP-morphism

Proof.

$$\begin{array}{ccc} \mathcal{O}_X & \xrightarrow{\rho} & \underline{\Omega}^0_{X/S} \\ \downarrow{\scriptstyle \simeq} & & \downarrow{\scriptstyle \phi^*} \\ R\phi_*\mathcal{O}_Y & \xrightarrow[R\phi_*\rho]{\simeq} & R\phi_*\Omega^0_{Y/S} \; (\simeq R\phi_*\underline{\Omega}^0_{Y/S}) \end{array}$$

□

Now (3.7) implies the following:

6.10 THEOREM. *Let $f : X \to S$ be a morphism of projective algebraic varieties of dimension n and k respectively, S an Abelian variety and $\mathcal{L}$ an ample line bundle on X. Then*

$$\mathbb{H}^p(X, \underline{\Omega}^q_{X/S} \otimes \mathcal{L}) = 0 \qquad p + q > n - k.$$

6.11 COROLLARY. *If in addition f is an SP-morphism, then*

$$H^p(X,\mathcal{L}) = 0 \qquad p > n - k.$$

This in turn implies that ω_X is not ample and that can be used to show that the fibres of $f : X \to S$ are isomorphic. Hence the fibres of g are birational.

§A. Appendix: The proof of (1.2.2)

The definition of $\mathbb{G}^p$ implies that there exists an exact sequence,

$$\dots \to R^{p+q}\Phi(\mathbb{F}^{p+1}) \xrightarrow{\alpha_{p,q}} R^{p+q}\Phi(\mathbb{F}^p) \xrightarrow{\beta_{p,q}} R^{p+q}\Phi(\mathbb{G}^p) \xrightarrow{\gamma_{p,q}} R^{p+q+1}\Phi(\mathbb{F}^{p+1}) \to \dots.$$

Therefore one has a morphism,

$$\beta_{p+1,q} \circ \gamma_{p,q} : R^{p+q}\Phi(\mathbb{G}^p) \to R^{p+q+1}\Phi(\mathbb{G}^{p+1}),$$

giving

$$d_1^{p,q} : E_1^{p,q} \to E_1^{p+1,q},$$

the first differential of E_r. Here $\gamma \circ \beta = 0$ shows that indeed $d^2 = 0$.

Next let

$$x \in \ker d_1^{p,q} \subseteq E_1^{p,q} = R^{p+q}\Phi(\mathbb{G}^p).$$

Since $\beta_{p+1,q}(\gamma_{p,q}(x)) = 0$, there exists a

$$y^1_{p,q}(x) \in R^{p+q+1}\Phi(\mathbb{F}^{p+2})$$

such that

$$\alpha_{p+1,q}(y^1_{p,q}(x)) = \gamma_{p,q}(x).$$

If

$$z^1_{p,q}(x) \in R^{p+q+1}\Phi(\mathbb{F}^{p+2})$$

is another element such that

$$\alpha_{p+1,q}(z^1_{p,q}(x)) = \gamma_{p,q}(x),$$

then there exists a

$$w \in R^{p+q}\Phi(\mathbb{G}^{p+1}) = E_1^{p+1,q-1}$$

such that

$$z^1_{p,q}(x) = y^1_{p,q}(x) + \gamma_{p+1,q-1}(w).$$

Thus

$$\beta_{p+2,q-1}(z^1_{p,q}(x)) = \beta_{p+2,q-1}(y^1_{p,q}(x)) + \underbrace{\beta_{p+2,q-1} \circ \gamma_{p+1,q-1}(w)}_{d_1^{p+1,q-1}(w)}.$$

Therefore

$$\beta_{p+2,q-1}(y^1_{p,q}(x)) + \operatorname{im} d_1^{p+1,q-1}$$

does not depend on the choice of $y^1_{p,q}(x)$. Fix such a $y^1_{p,q}(x)$ for each x. Then one has a well-defined natural morphism,

$$d_2^{p,q} : E_2^{p,q} = \ker d_1^{p,q} \Big/ \operatorname{im} d_1^{p-1,q} \to E_2^{p+2,q-1} = \ker d_1^{p+2,q-1} \Big/ \operatorname{im} d_1^{p+1,q-1}.$$

Now let

$$K_2^{p,q} = \{u \in \ker d_1^{p,q} \mid u + \operatorname{im} d_1^{p-1,q} \in \ker d_2^{p,q}\}$$

and let $x \in K_2^{p,q}$. Then there exists a

$$y^1_{p,q}(x) \in R^{p+q+1}\Phi(\mathbb{F}^{p+2})$$

such that

$$\alpha_{p+1,q}(y^1_{p,q}(x)) = \gamma_{p,q}(x) \tag{A.1}$$
$$\beta_{p+2,q-1}(y^1_{p,q}(x)) = \beta_{p+2,q-1} \circ \gamma_{p+1,q-1}(x')$$

for some $x' \in R^{p+q}\Phi(\mathbb{G}^{p+1})$, and so

$$y^1_{p,q}(x) - \gamma_{p+1,q-1}(x') \in \ker \beta_{p+2,q-1} = \operatorname{im} \alpha_{p+2,q-1}.$$

Then there exists a

$$y^2_{p,q}(x) \in R^{p+q+1}\Phi(\mathbb{F}^{p+3})$$

such that

$$\alpha_{p+2,q-1}(y^2_{p,q}(x)) = y^1_{p,q}(x) - \gamma_{p+1,q-1}(x'). \tag{A.2}$$

Now

$$\beta_{p+3,q-2}(y^2_{p,q}(x)) \in \ker d_1^{p+3,q-2}$$

and the class,

$$\beta_{p+3,q-2}(y^2_{p,q}(x)) + \operatorname{im} d_1^{p+2,q-2} \in \ker d_1^{p+3,q-2} \Big/ \operatorname{im} d_1^{p+2,q-2} = E_2^{p+3,q-2}$$

does not depend on the choice of $y^2_{p,q}(x)$ as long as it satisfies $(A.2)$. Furthermore, since $\gamma \circ \beta = 0$,

$$\beta_{p+3,q-2}(y^2_{p,q}(x)) + \operatorname{im} d_1^{p+2,q-2} \in \ker d_2^{p+3,q-2}.$$

Thus one can define

$$d_3^{p,q} : E_3^{p,q} = \ker d_2^{p,q} \Big/ \operatorname{im} d_2^{p-2,q+1} \to E_3^{p+3,q-2} = \ker d_2^{p+3,q-2} \Big/ \operatorname{im} d_2^{p+1,q-1}$$

by the formula

$$d_3^{p,q}\left(x + \operatorname{im} d_1^{p-1,q} + \operatorname{im} d_2^{p-2,q+1}\right) = \beta_{p+3,q-2}(y^2_{p,q}(x)) + \operatorname{im} d_1^{p+2,q-2} + \operatorname{im} d_2^{p+1,q-1}.$$

Claim. $d_3^{p,q}$ is well-defined.

Proof. One needs to prove that $d_3^{p,q}$ does not depend on the choice of $y^1_{p,q}(x)$ and x'. Let $z^1_{p,q}(x) \in R^{p+q+1}\Phi(\mathbb{F}^{p+2})$ and $z^2_{p,q}(x) \in R^{p+q+1}\Phi(\mathbb{F}^{p+3})$ be such that

$$\alpha_{p+1,q}(z^1_{p,q}(x)) = \gamma_{p,q}(x) \tag{A.3}$$
$$\beta_{p+2,q-1}(z^1_{p,q}(x)) = \beta_{p+2,q-1} \circ \gamma_{p+1,q-1}(x'')$$

for some $x'' \in R^{p+q}\Phi(\mathbb{G}^{p+1})$ and

$$\alpha_{p+2,q-1}(z^2_{p,q}(x)) = z^1_{p,q}(x) - \gamma_{p+1,q-1}(x'').$$

By $(A.1)$ and $(A.3)$ there exists an $x''' \in R^{p+q}\Phi(\mathbb{G}^{p+1})$ such that

$$z^1_{p,q}(x) - \gamma_{p+1,q-1}(x'') = y^1_{p,q}(x) - \gamma_{p+1,q-1}(x') + \gamma_{p+1,q-1}(x''').$$

Then

$$\gamma_{p+1,q-1}(x''') = \alpha_{p+2,q-1}(z^2_{p,q}(x) - y^2_{p,q}(x)),$$

hence by definition,

$$d_2^{p+1,q-1}(x''' + \operatorname{im} d_1^{p,q-1}) = \beta_{p+3,q-2}(z^2_{p,q}(x) - y^2_{p,q}(x)) + \operatorname{im} d_1^{p+2,q-2}.$$

In other words

$$\beta_{p+3,q-2}(z^2_{p,q}(x) - y^2_{p,q}(x)) + \operatorname{im} d_1^{p+2,q-2} \in \operatorname{im} d_2^{p+1,q-1}$$

which means that $d_3^{p,q}$ is indeed well-defined. □

In the general case one has to work in a similar way. Assume that for a fixed $r \in \mathbb{N}$ and for all p, q and $1 \leq i \leq r$,

$$d_i^{p,q} : E_i^{p,q} \to E_i^{p+i,q-i+1}$$

is defined and

$$E_i^{p,q} \simeq \ker d_{i-1}^{p,q} \big/ \operatorname{im} d_{i-1}^{p-i+1,q+i-2}.$$

Let $K_0^{p,q} = E_1^{p,q}$ and

$$K_i^{p,q} = \{x \in E_1^{p,q} \mid x \in \ker d_1, x + \operatorname{im} d_1 \in \ker d_2, x + \operatorname{im} d_1 + \operatorname{im} d_2 \in \ker d_3, \dots$$
$$\dots, x + \operatorname{im} d_1 + \operatorname{im} d_2 + \cdots + \operatorname{im} d_{i-1} \in \ker d_i\}.$$

Then $K_{i-1}^{p,q}$ maps surjectively onto $E_i^{p,q}$. Let the kernel of this surjection be $I_{i-1}^{p,q}$, i.e.,

$$E_i^{p,q} \simeq K_{i-1}^{p,q} \big/ I_{i-1}^{p,q}.$$

Then $I_{i-1}^{p,q}$ is the preimage of $\operatorname{im} d_{i-1}^{p-i+1,q+i-2}$ in $K_{i-1}^{p,q}$. Let

$$\alpha^i_{p,q} = \alpha_{p+1,q} \circ \alpha_{p+2,q-1} \circ \cdots \circ \alpha_{p+i,q-i+1}.$$

Further assume that there exist (set-theoretic) functions,

$$y^{i-1}_{p,q} : K_{i-1}^{p,q} \to R^{p+q+1}\Phi(\mathbb{F}^{p+i})$$

such that

$$\alpha^{i-1}_{p,q}(y^{i-1}_{p,q}(x)) = \gamma_{p,q}(x) \in R^{p+q+1}\Phi(\mathbb{F}^{p+1}). \tag{A.4}$$

Further notice, that

$$d_i^{p,q} : E_i^{p,q} \to E_i^{p+i,q-i+1}$$

is given by

$$d_i^{p,q}\left(x + I_{i-1}^{p,q}\right) = \beta_{p+i,q-i+1}(y^{i-1}_{p,q}(x)) + I_{i-1}^{p+i,q-i+1} \tag{A.5}$$

and $d_i^{p,q}$ does not depend on the choice of $y^j_{p,q}(x)$ as long as it satisfies $(A.4)$.

Note also that $y^0_{p,q}(x) = \gamma_{p,q}(x)$. Let $I_r^{p,q}$ be the preimage of $\operatorname{im} d_r^{p-r,q+r-1}$ in $K_r^{p,q}$ and let

$$E_{r+1}^{p,q} \simeq K_r^{p,q} / I_r^{p,q}.$$

Suppose now that $x \in K_r^{p,q}$, i.e., $x + I_{r-1}^{p,q} \in \ker d_r^{p,q}$ or equivalently

$$\beta_{p+r,q-r+1}(y^{r-1}_{p,q}(x)) \in I_{r-1}^{p+r,q-r+1}.$$

By $(A.5)$ there exists an $x_j \in K_j^{p+r-j-1,q-r+j+1}$ for $j = 0, \dots, r-2$ such that

$$\beta_{p+r,q-r+1}\Big(y^{r-1}_{p,q}(x) - \sum_{j=0}^{r-2} y^j_{p+r-j-1,q-r+j+1}(x_j)\Big) = 0.$$

Therefore there exists a

$$y^r_{p,q}(x) \in R^{p+q+1}\Phi(\mathbb{F}^{p+r+1})$$

such that

$$\alpha_{p+r,q-r+1}(y^r_{p,q}(x)) = y^{r-1}_{p,q}(x) - \sum_{j=0}^{r-2} y^j_{p+r-j-1,q-r+j+1}(x_j).$$

By $(A.4)$, and since $\alpha \circ \gamma = 0$,

$$\begin{aligned} \alpha^r_{p,q}(y^r_{p,q}(x)) &= \alpha^{r-1}_{p,q} \circ \alpha_{p+r,q-r+1}(y^r_{p,q}(x)) \\ &= \alpha^{r-1}_{p,q}\Big(y^{r-1}_{p,q}(x) - \sum_{j=0}^{r-2} y^j_{p+r-j-1,q-r+j+1}(x_j)\Big) \\ &= \alpha^{r-1}_{p,q}(y^{r-1}_{p,q}(x)) = \gamma_{p,q}(x). \end{aligned} \tag{A.6}$$

Now one can define

$$d^{p,q}_{r+1} : E^{p,q}_{r+1} = K^{p,q}_r/I^{p,q}_r \to E^{p+r+1,q-r}_{r+1} = K^{p+r+1,q-r}_r \Big/ I^{p+r+1,q-r}_r$$

by

$$d^{p,q}_{r+1}\,(x + I^{p,q}_r) = \beta_{p+r+1,q-r}(y^r_{p,q}(x)) + I^{p+r+1,q-r}_r.$$

Claim. $d^{p,q}_{r+1}$ is well-defined.

Proof. One needs to prove that any set of functions $z^i_{p,q}$ for p, q and $0 \leq i \leq r$ satisfying $(A.4)$, $(A.5)$ and $(A.6)$ gives the same $d^{p,q}_{r+1}$. This is true for $d^{p,q}_i$, $1 \leq i \leq r$, by induction.

Since

$$\alpha_{p+1,q} \circ \alpha^{i-1}_{p+1,q-1}(z^i_{p,q}(x)) = \alpha^i_{p,q}(z^i_{p,q}(x)) = \gamma_{p,q}(x),$$

one finds that

$$\alpha^i_{p+1,q-1}(z^i_{p,q}(x)) = z^1_{p,q}(x) + \gamma_{p+1,q-1}(x_i)$$

for some $x_i \in E^{p+1,q-1}_1$ for all $i \geq 2$. Now replace $z^i_{p,q}(x)$ by $z^i_{p,q}(x) - z^{i-1}_{p+1,q-1}(x_i)$ for $i \geq 2$. Then $\alpha^i_{p,q}(z^i_{p,q}(x)) = \gamma_{p,q}(x)$ remains true and $\beta_{p+i+1,q-i}(z^i_{p,q}(x))$ is only changed by an element of $I^{p+i+1,q-i}_i$. Hence one may assume that

$$\alpha^i_{p+1,q-1}(z^i_{p,q}(x)) = z^1_{p,q}(x).$$

Repeating this argument for $j = 2, \ldots, r-1$, using $\alpha^{i-j}_{p+j,q-j}$ in place of $\alpha^{i-1}_{p+1,q-1}$, one may assume that

$$\alpha^{i-j}_{p+j,q-j}(z^i_{p,q}(x)) = z^j_{p,q}(x) \quad \text{for all } i > j.$$

One may also assume the same for the y's, i.e., for all p, q and $0 \leq j \leq r-1$,

$$\alpha^{i-j}_{p+j,q-j}(y^i_{p,q}(x)) = y^j_{p,q}(x) \quad \text{for all } i > j. \tag{A.7}$$

Henceforth it will be assumed that any set of y's satisfies $(A.7)$. Observe that

$$\alpha_{p+1,q}(z^1_{p,q}(x)) = \gamma_{p,q}(x) = \alpha_{p+1,q}(y^1_{p,q}(x)),$$

so there exists an $x' \in E_1^{p+1,q-1}$ such that

$$z^1_{p,q}(x) = y^1_{p,q}(x) + \gamma_{p+1,q-1}(x').$$

Then replace $y^i_{p,q}(x)$ by $y^i_{p,q}(x) + y^{i-1}_{p+1,q-1}(x')$ for $i \geq 1$. This way

$$\alpha^i_{p,q}(y^i_{p,q}(x)) = \gamma_{p,q}(x)$$

and $(A.7)$ remain true and $\beta_{p+i+1,q-i}(y^i_{p,q}(x))$ is only changed by an element of $I_i^{p+i+1,q-i}$. Hence one may assume that

$$z^1_{p,q}(x) = y^1_{p,q}(x).$$

Next observe that

$$\alpha_{p+2,q-1}(z^2_{p,q}(x)) = z^1_{p,q}(x) = y^1_{p,q}(x) = \alpha_{p+2,q-1}(y^2_{p,q}(x)),$$

hence there exists an $x'' \in E_1^{p+2,q-2}$ such that

$$z^2_{p,q}(x) = y^2_{p,q}(x) + \gamma_{p+2,q-2}(x'').$$

Now replace $y^i_{p,q}(x)$ by $y^i_{p,q}(x) + y^{i-2}_{p+2,q-2}(x'')$ for $i \geq 2$. Again

$$\alpha^i_{p,q}(y^i_{p,q}(x)) = \gamma_{p,q}(x)$$

and $(A.7)$ remain true and $\beta_{p+i+1,q-i}(y^i_{p,q}(x))$ is only changed by an element of $I_i^{p+i+1,q-i}$. Hence one can assume that

$$z^2_{p,q}(x) = y^2_{p,q}(x)$$

also.

Iterating this procedure one finds that $d^{p,q}_{r+1}$ is the same defined either via the y's or the z's. □

Now let $z \in R^{p+q}\Phi(\mathbb{F}^p)$. Then $\beta_{p,q}(z) \in K_i^{p,q}$ for all i, since $\gamma \circ \beta = 0$. On the other hand if $x \in K^{p,q}_{k-p}$, then $y^{k-p}_{p,q}(x) = 0$, since $\mathbb{F}^{k+1} = 0$. Then

$$\gamma_{p,q}(x) = \alpha^{k-p}_{p,q}(y^{k-p}_{p,q}(x)) = 0,$$

so there exists a $z \in R^{p+q}\Phi(\mathbb{F}^p)$ such that $x = \beta_{p,q}(z)$.

Therefore $\beta_{p,q}$ induces a surjective morphism

$$\bar{\beta}_{p,q} : R^{p+q}\Phi(\mathbb{F}^p) \twoheadrightarrow E^{p,q}_\infty.$$

Let

$$F^pR^{p+q} = \alpha^{p-l}_{l-1,p+q-l}(R^{p+q}\Phi(\mathbb{F}^p)) \subseteq R^{p+q}\Phi(\mathbb{F}^l).$$

From the construction of the spectral sequence it is clear that $\bar{\beta}_{p,q}(x) = 0$ for any $x \in R^{p+q}\Phi(\mathbb{F}^p)$ such that $\alpha^{p-l}_{l-1,p+q-l}(x) = 0$, hence $\bar{\beta}_{p,q}$ factors through F^pR^{p+q} inducing a surjective morphism

$$\hat{\beta}_{p,q} : F^pR^{p+q} \twoheadrightarrow E^{p,q}_\infty.$$

Now $\ker \beta_{p,q} = \operatorname{im} \alpha_{p,q}$ implies that $\ker \hat{\beta}_{p,q} = F^{p+1}R^{p+q}$. Hence for any m

$$R^m\Phi(\mathbb{F}^l) = F^lR^m \supseteq F^{l+1}R^m \supseteq \cdots \supseteq F^kR^m \supseteq F^{k+1}R^m = 0$$

is a filtration of $R^m\Phi(K) = R^m\Phi(\mathbb{F}^l)$ such that

$$E_\infty^{p,m-p} \simeq F^pR^m \big/ F^{p+1}R^m.$$

This completes the proof of (1.2.2). □

References

[CKM88] H. Clemens, J. Kollár, S. Mori, *Higher Dimensional Complex Geometry*, vol. 166, Astérisque, 1988.

[Hartshorne66] R. Hartshorne, *Residues and Duality*, Lecture Notes in Math, vol. 20, Springer-Verlag, 1966.

[Fulton84] W. Fulton, *Intersection Theory*, Springer, 1984.

[Kovács96] S. J. Kovács, *Smooth families over rational and elliptic curves*, J. Alg. Geom., **5** (1996), 369–385.

[Kovács97a] S. J. Kovács, *Families over a base with a birationally nef tangent bundle*, Math. Ann. **308** (1997), 347–359.

[Kovács97b] S. J. Kovács, *Relative De Rham complexes for non-smooth morphisms*, Birational Algebraic Geometry, Y. Kawamata, V. Shokurov eds., Contemp. Math., vol. 207, 1997, pp. 89–100.

[Kovács02] S. J. Kovács, *Logarithmic vanishing theorems and Arakelov-Parshin boundedness for singular varieties*, Compositio Math. **131** (2002), 291–317.

[Migliorini95] L. Migliorini, *A smooth family of minimal surfaces of general type over a curve of genus at most one is trivial*, J. Alg. Geom. **4** (1995), 353–361.

University of Washington, Department of Mathematics, Seattle, WA 98195
E-mail address: kovacs@math.washington.edu

Contemporary Mathematics
Volume **386**, 2005

Holomorphic forms on threefolds

Tie Luo and Qi Zhang

ABSTRACT. Two conjectures relating the Kodaira dimension of a smooth projective variety and the number of linearly independent nowhere vanishing 1-forms on the variety are proposed and verified in dimension 3.

§1 introduction

Let X be a smooth projective variety over the complex number field $\mathbb{C}$. Let $\omega \in H^0(X, \Omega^i_X)$ be a nontrivial global holomorphic differential i-form. We would like to understand the nature of zero locus of ω in terms of the birational geometry of X. On the one hand algebraic varieties are birationally classified according to their Kodaira dimensions whose definition uses essentially high multiples of holomorphic form of top degree-the canonical bundle and Mori theory of extremal contractions and flips are birational operations done altering the base loci of (multiples of) the canonical bundle, on the other hand very little is known about how much impact the zero locus of a holomorphic form of lower degree has on the birational geometry of the underlying variety. The first question one may ask is: What makes ω to have zero locus? It is proved in [Z] that for any global holomorphic 1-form $0 \neq \omega \in H^0(X, \Omega^1_X)$, its zero locus $Z(\omega)$ is not empty provided that the canonical bundle K_X is ample. It is natural to suspect that the same conclusion should hold for varieties of general type.

In the following we propose two conjectures (Conjectures 1 and 2). Indeed, Conjecture 2 is a stronger form of Conjecture 1.

CONJECTURE 1. Assume X is of general type and $0 \neq \omega \in H^0(X, \Omega^1_X)$. Then $Z(\omega) \neq \emptyset$.

and more precisely

CONJECTURE 2. Assume there exists a subspace V in $H^0(X, \Omega^1_X)$ of dimension k; $1 \leq k \leq \dim(X)$, such that $Z(\omega) = \phi$ holds for any $0 \neq \omega \in V$. Then the Kodaira dimension of X is less than or equal to $\dim(X) - k$.

2000 *Mathematics Subject Classification.* 14E30 14F10.

These conjectures are trivially true for curves. The proofs of these conjectures are also straight forward for surfaces.

The purpose of the present paper is to verify these conjectures in dimension 3. Namely, we prove the following Theorems 1 and 2.

THEOREM 1. *Let X be a smooth projective threefold of general type. For any global holomorphic 1-form $0 \neq \omega \in H^0(X, \Omega_X^1)$, its zero locus $Z(\omega)$ is not empty.*

and

THEOREM 2. *Let X be a smooth projective threefold admitting a k-dimensional subspace $V \subset H^0(X, \Omega_X^1)$ such that $Z(\omega) = \emptyset$ for any $0 \neq \omega \in V$. Then the Kodaira dimension of X is at most $3 - k$.*

Some of the techniques used in proving these results can not be generalized to higher dimensions. We plan to discuss similar results in higher dimensions using different tools in a future paper.

In contrast to the result on 1-forms, there are examples (see section 4 below) of threefolds with ample canonical bundle and carry nowhere vanishing 2-forms. Using a completely different approach and viewing a 2-form as a meromorphic vector field, we have

THEOREM 3. *Let X be a threefold of general type. $\omega \in H^0(X, K_X \otimes \Omega_X^2)$ is a canonically twisted holomorphic 2-form. Then the zero locus $Z(\omega)$ of ω is not empty.*

ACKNOWLEDGMENTS. We thank S. Keel, J. McKernan, and M. Reid for their interest in and comments on our results. The referee's comments also are greatly appreciated.

§2 Proof of Theorem 1

The following result uses the classification of extremal contractions in dimension 3. It is related to the ideas used in [L] and reduces the proof of Theorem 1 to those X admitting smooth minimal models.

LEMMA 2.1. *Assume X is a smooth projective threefold of general type. If X does not admit a smooth minimal model, then every holomorphic 1-form on X has nonempty zero locus.*

PROOF. Since K_X is not nef, there is an extremal contraction $\pi : X \to Y$. Let E be the exceptional locus. The classification in [M] says that f is divisorial with the exceptional divisor $\mathbb{P}^2$ with normal bundle $\mathcal{O}(-1)$ or $\mathcal{O}(-2)$, $\mathbb{P}^1 \times \mathbb{P}^1$ with normal bundle $\mathcal{O}(-1) \otimes \mathcal{O}(-1)$, a cone E over rational normal curve of degree 2 with normal bundle $\mathcal{O}(-1)$, or a $\mathbb{P}^1$ bundle over a smooth curve.

One has the following exact sequence

$$0 \to N_{E/X}^* \to \Omega_X^1|_E \to \Omega_E^1 \to 0$$

where $N_{E/X}^*$ denotes the conormal bundle of E in X.

Note that unless E is a $\mathbb{P}^1$ bundle over a smooth curve, $H^0(E, \Omega_E^1) = 0$. This implies that restrictions of a nontrivial global 1-form on E provides a nontrivial global section of positive line bundle $N_{E/X}^*$, which has nontrivial zero locus.

When E is a $\mathbb{P}^1$ bundle over a smooth curve, the resulting Y is smooth. From our assumption, K_Y is not nef. One can repeat the argument done for X and the process has to terminate after finitely many steps. So every holomorphic 1-form on X has nonempty zero locus.

□

Remark. Indeed the proof shows that if X does not admits a smooth minimal model, a 1-form has at least one dimensional zero locus.

Due to Lemma 2.1, in order to prove Theorem 1, we may assume that the canonical bundle of X is nef and big. One possible way of proving Theorem 1 would be the following (see [R1] for related technical terms). Let $f : X \to Y$ be the birational morphism to the canonical model of X. Y is known to have Gorenstein canonical singularities. One may assume that Y is not terminal. Reid [R1] constructed explicitly a crepant resolution for such singularities. Let $g : X' \to Y$ be the resolution such that g is crepant and X' has at most isolated cDV points. One may assume, after further small resolution, that X' is $\mathbb{Q}$-factorial. Since both K_X and $K_{X'}$ are nef, it is known that X and X' are transformed to each other by applying a sequence of flops. Thus X' is smooth. One way to prove Theorem 1 is to show that any 1-form on X' has to have nontrivial zero locus (which appears in the process of explicitly resolving the singularities on Y) and flops do not change the existence of the zero locus.

Here we take a different approach using Kodaira-Akizuki-Nakano type vanishing theorem for semismall morphisms proved by Migliorini as in [Mi]. A birational morphism ϕ is called *semismall* if for every irreducible component E in the exceptional locus,

$$\text{codim}E \geq \dim\phi^{-1}(p)$$

for any $p \in \phi(E)$. For example in dimension 3 a semismall morphism contracts curves to points or divisor to curves.

PROPOSITION 2.2(THEOREM 4.6 IN [MI]). *Let X be a smooth projective variety over $\mathbb{C}$ with a semiample divisor L, namely some multiple of L defines a birational morphism. Assume the morphism $\phi_{|mL|}$ is semismall for some $m > 0$. Then*

$$H^p(X, L^{-1} \otimes \Omega^q_X) = 0$$

for $p + q < \dim X$.

The following general result deals with maps which contract divisors to points.

LEMMA 2.3. *Let $\pi : X \to Y$ be a projective birational morphism with X smooth and Y normal. Assume an irreducible divisor $E \subset X$ is contracted to a point. Let $\omega \in H^0(X, \Omega^1_X)$. Then $Z(\omega) \cap E \neq \emptyset$.*

PROOF. Assume $\pi(E) = p$. Let us embed a neighborhood of p in some $\mathbb{C}^N$ so that p is mapped to the origin O and still call the composition π. For any $q \in E$, one has $\pi^* : \mathfrak{m}_{O,\mathbb{C}^N}/\mathfrak{m}^2_{O,\mathbb{C}^N} \to \mathfrak{m}_{q,X}/\mathfrak{m}^2_{q,X}$ between the Zariski cotangent spaces.

If q is a singularity of E and assume E is locally defined by some f, then $f \in \mathfrak{m}^2_{q,X}$. This implies $\pi^* = 0$ and $q \in Z(\omega)$. If E is smooth at q, then the rank of

π^* is either 0 or 1. When the rank is 0, again $q \in Z(\omega)$. So we may reduce the proof of the desired result to the case in which the rank of π^* is 1 along E (in particular E is smooth and π^* of any element of $\mathfrak{m}_{O,\mathbb{C}^N}/\mathfrak{m}^2_{O,\mathbb{C}^N}$ is a scalar multiple of the class $x_1 + \mathfrak{m}^2_{q,X}$ where x_1 locally defines E. The map $\mathfrak{m}_{q,X}/\mathfrak{m}^2_{q,X} \to \mathfrak{m}_{q,E}/\mathfrak{m}^2_{q,E}$ sends a scalar multiple of the class $x_1 + \mathfrak{m}^2_{Q,X}$ to 0).

In this case ω defines a nowhere vanishing section of $N^*_{E/X}$. This contradicts the adjuction lemma of Kawamata in [K].

□

Now we can finish the proof of Theorem 1.

THEOREM 2.4=THEOREM 1. *Let X be a threefold of general type. For any global holomorphic 1-form $0 \neq \omega \in H^0(X, \Omega^1_X)$, its zero locus $Z(\omega)$ is not empty.*

PROOF. If K_X is not nef, Lemma 2.1 says $Z(\omega)$ is not empty. Assume K_X is nef, then K_X is semiample. Let $\phi_m = \phi_{mK_X}$ for $m \gg 0$. By Lemma 2.3 we may assume ϕ_m is semismall. Assuming $Z(\omega)$ is empty, one has exact sequence

$$0 \to \mathcal{O} \xrightarrow{\omega} \Omega^1_X \xrightarrow{\wedge\omega} \Omega^2_X \xrightarrow{\wedge\omega} \Omega^3_X \to 0.$$

Arguing as in [Z], using Proposition 2.2 for $L = K_X$ one gets an embedding of Ω^3_X into Ω^2_X. It follows that $H^0(X, K_X^{-1} \otimes \Omega^2_X) \neq 0$ which is impossible again by 2.2.

□

It is easy to construct a threefold of general type with ample canonical bundle admitting a 1-form whose zero locus is of dimension 0, 1, or 2.

§3 Proof of Theorem 2

On a threefold X, the maximal dimension of a subspace $V \subset H^0(X, \Omega^1_X)$ with the property that any $0 \neq \omega \in V$ has no empty locus is 3. This is due to the fact that the rank of Ω^1_X is 3. By taking a product of a variety of general type and an Abelian variety, one gets examples of threefold of Kodaira dimension $3 - k$ and having a k-dimensional subspace of 1-forms with the property that any nonzero element is nowhere zero.

THEOREM 3.1=THEOREM 2. *Let X be a threefold admitting a k-dimensional subspace $V \subset H^0(X, \Omega^1_X)$ such that $Z(\omega) = \emptyset$ for any $0 \neq \omega \in V$. Then the Kodaira dimension of X is at most $3 - k$.*

PROOF. When $k = 1$, the claim is implied by Theorem 1. When $k = 3$, let $\omega_1, \omega_2, \omega_3$ be a basis of V, $\omega_1 \wedge \omega_2 \wedge \omega_3$ provides a nowhere zero section of K_X. So K_X is trivial and $\kappa(X) = 0$.

Let us assume $k = 2$. By Lemma 2.1 we see that X is minimal because of existence of V. We also know $\kappa(X) \neq 3$ by Theorem 1. Thus it is enough to show $\kappa(X) \neq 2$. Assuming the contrary holds. Let $\phi : X \to S$ be the Iitaka fibration defined by high multiples of K_X. $\phi^{-1}(s)$ is an elliptic curve for a general $s \in S$. S is surface of general type with at worst canonical singularities. Let $\alpha : X \to \mathrm{Alb}(X)$

be the Albanese map. The existence of V implies that $\alpha(X)$ is at least of dimension 2.

We consider two cases:

Case I. $\alpha(\phi^{-1}(s))$ is a point for a general $s \in S$. In this case $\alpha(X)$ has dimension 2. Let $X \to Y \to \alpha(X)$ be the Stein factorization of α. It is easy to see that S and Y are birationally equivalent. In particular 1-forms on X are pullbacks of 1-forms (on the smooth locus) of S. Assume the subspace V has a basis ω_1, ω_2, which correspond to ω_1', ω_2' on S. ω_1', ω_2' span a 2-dimensional subspace V' of $H^0(S, \Omega_S^1)$ having the property that for any $0 \neq \omega' \in V'$ the zero locus on the smooth locus of S is empty. Then $\omega_1' \wedge \omega_2'$ generates a nowhere zero section of K_S on the smooth locus. Hence $\kappa(S) = 0$, a contradiction.

Case II. $\alpha(\phi^{-1}(s))$ is a curve for a general $s \in S$. In this case $0 = \kappa(\phi^{-1}(s)) \geq \kappa(\alpha(\phi^{-1}(s)) \geq 0$. So $\alpha(\phi^{-1}(s)) = E_s$ is also an elliptic curve in $A = \mathrm{Alb}(X)$. Since A contains at most countably many sub Abelian varieties of dimension 1, E_s are translations of a particular sub Abelian variety A_1 of dimension 1 . Let $\pi : A \to A/A_1$ be the quotient map. One has the following diagram:

$$\begin{array}{ccc} X & \xrightarrow{\alpha} & A \supset \alpha(X) \\ {\scriptstyle\phi}\downarrow & & \downarrow{\scriptstyle\pi} \\ S & & A/A_1 \supset \pi\alpha(X) \end{array}$$

If $\alpha(X)$ is of dimension 3, $\pi(\alpha(X))$ is of dimension 2. Again let $X \to Y \to \pi(\alpha(X))$ be the Stein factorization of $\pi \circ \alpha$. S is birationally equivalent to Y and we are done by case I. If $\alpha(X)$ is of dimension 2, $C = \pi(\alpha(X))$ is of dimension 1 in the Abelian variety A/A_1. Let ω_1', ω_2' be the 1-forms on A which pullback to a basis of V. There is a nontrivial linear combination $\omega' = c_1\omega_1' + c_2\omega_2'$ which is identically zero on the fibers of π. ω' is a pullback of some ω'' on A/A_1. Assume C is of general type and we may assume C is smooth, $\omega"|C$ vanishes at some point $p \in C$, which in turn implies $c_1\omega_1 + c_2\omega_2$ vanishes on the fiber of $\pi \circ \alpha$ in X over p. If C is a smooth elliptic curve, the pullback of $\omega' = c_1\omega_1' + c_2\omega_2'$ (which is the same as the pullback of ω'') would have a zero unless the morphism $\pi \circ \alpha$ is smooth and ω'' does not vanish on C. So let us assume $\pi \circ \alpha$ is smooth. Note that each fiber the morphism has to be a surface of general type in order to keeping $\kappa(X) = 2$. Furthermore each fiber has to be minimal since K_X is nef. Thus $\pi \circ \alpha : X \to C$ represents a smooth family of minimal surfaces of general type over an elliptic curve. It has to be isotrivial thanks to Theorem 1.1 in [Mi]. So up to an étale covering, X is isomorphic to the product of C and a fiber of $\pi \circ \alpha$. It implies that V contains a nonzero member ω coming from a 1-form on the fiber which is a surface of general type. As any 1-form on a surface of general type has to have a zero, $Z(\omega)$ contains a curve. A contradiction.

□

§4 2-forms on threefolds of general type

After the discussion regarding zeros of 1-forms, it is quite natural to wonder whether similar results on 1-forms still hold true for higher orders forms, i.e;

Question: Let X be a threefold of general type. Is it true that any global differential 2-form on X must have nonempty zero locus?

Campana and Peternell [CP] constructed the following example which shows that a global section of Ω_X^2 may not have zero locus even when K_X is ample:

EXAMPLE. Let Y be a projective holomorphic symplectic fourfold. Let X be a hypersurface of sufficiently high degree. Then the nondegenerate 2-form on Y provides a nowhere vanishing 2-form on X.

However by virtue of [L] it is known that when K_X is not nef, any 2-form has nontrivial zero locus. It would be interesting to see whether there are examples with nef canonical bundle and a 2-form whose zero locus is empty.

Here we will show that any canonically twisted 2-form $\omega \in H^0(X, K_X \otimes \Omega_X^2)$ on a threefold has nontrivial zero locus.

Let X be a smooth projective variety of dimension n. Let $\theta \in H^0(X, T_X \otimes L)$ be an element which has isolated zeros. Then locally around each zero p, under coordinates $z_1, z_2, ..., z_n$, one can express θ as

$$\theta = \sum_{i=1}^{n} f_i s \otimes \frac{\partial}{\partial z_i}$$

where f_i are holomorphic functions and s is a local section of L. The Jacobian at the zero p is defined as

$$A_p = (\frac{\partial f_i}{\partial z_j})_{|p}$$

which is an $n \times n$ matrix. A_p is invariant under coordinate change upto a nowhere zero holomorphic function. p is nondegenerate if $det(A_p) \neq 0$.

The following lemma is a corollary of the residue theorem due to Bott (see [C]).

LEMMA 4.1. *Notations as above and assume θ has only nondegenerate zeros. Then*

$$\int_X \bar{c}_1(T_X \otimes L)^n = \sum_{p \in Z(\theta)} \frac{(Tr(A_p))^n}{det(A_p)}$$

and

$$\int_X \bar{c}_n(T_X \otimes L) = |Z(\theta)|$$

where $\bar{c}_r(T_X \otimes L) = \sum_{i=0}^{r} c_{r-i}(T_X) c_1(L)^i$.

Remark. Indeed Chern's proof [C] shows that away from the zero locus, $\bar{c}_1(T_X \otimes L)^n$ and $\bar{c}_n(T_X \otimes L)$ can be written as derived differential forms, namely they are of the forms d of differential forms of lower degrees. This fact is used in the proof of the following

THEOREM 4.2=THEOREM 3. *Let X be a threefold of general type. $\omega \in H^0(X, K_X \otimes \Omega_X^2)$ is a canonically twisted holomorphic 2-form. Then the zero locus $Z(\omega)$ of ω is not empty.*

PROOF. We may assume K_X is nef since otherwise there is a divisorial extremal contraction and an analysis similar to that in section 2 of [L] shows that for all possible divisorial extremal contraction any canonically twisted 2-form has nontrivial zero locus along the exceptional divisor.

Since Ω_X^2 is isomorphic to $K_X \otimes T_X$, ω corresponds to a global section θ in $H^0(X, 2K_X \otimes T_X)$, which can be viewed as a meromorphic vector field with coefficient in the line bundle $2K_X$. Assume θ (or equivalently ω) does not have zero, the remark about the above lemma says

$$\bar{c}_1(2K_X \otimes T_X)^3 = 0.$$

However the left hand side is

$$\bar{c}_1(2K_X \otimes T_X)^3 = (c_1(T_X) + c_1(2K_X))^3 = K_X^3 > 0,$$

a contradiction.

□

Finally the geometric meaning of $c_3(\Omega_X^2)$ (which is $c_3 - c_1c_2$ of X) in terms of singularities appearing on a minimal model is explained in:

THEOREM 4.3. *Let X be a threefold of general type. Assume Ω_X^2 admits a section ω with isolated nondegenerate zeros. Then $c_3(\Omega_X^2)$ is greater or equal to the number of singularities whose indices are bigger than 1 on a minimal model.*

PROOF. Let X_{min} be a minimal model of X. Assuming ω is a 2-form with only isolated nondegenerate zeros on X. By [L], ω corresponds to a (global locally invariant) 2-form on X_{min} not vanishing at the singularities whose indices are bigger than 1. Moreover every such singularity produces at least one isolated zero on X. Bott's residue formula, when taking the top Chern class, says

$$\bar{c}_3(K_X \otimes T_X) = \sum_{p \in Z(\omega)} \frac{det(A_p)}{det(A_p)} = \text{number of zeros of } \omega.$$

This implies

$$\bar{c}_3(K_X \otimes T_X) = c_3(T_X) + c_2(T_X)c_1(K_X) + c_1(T_X)c_1(K)^2 + c_1(K_X)^3 = c_3 - c_1c_2$$
$$\geq \text{number of singularities whose indices are bigger than 1.}$$

□

References

[C] Chern, S.S, *Meromorphic vector fields and characteristic numbers*, Scripta Mathematica **29** (1973), 243-251.

[CP] Campana, F and Peternell, T, *Holomorphic 2-forms on complex threefolds*, Journal of Algebraic Geometry **9** (2000), 223-264.

[K] Kawamata, Y, *On the length of an extremal rational curve*, Invent. Math. **105** (1991), 609-611.

[L] Luo, T, *Global 2-forms and pluricanonical systems on threefolds*, Math. Ann. **318** (2000), 707-730.

[Mi] Migliorini, L, *A smooth family of minimal surfaces of general type over a curve of genus at most one is trivial*, Journal of Algebraic Geometry **4** (1995), 356-363.

[Mo] Mori, S, *Threefolds whose canonical bundles are not numerically effective*, Ann. of Math. **116** (1982), 133-176.
[R1] Reid, M, *Canonical 3-folds*, Journée de Géometri algébrique d'Angers, 1980, pp. 273-310.
[R2] Reid, M, *Minimal models of canonical 3-folds*, Algebraic Varieties and Analytic Varieties, Adv. Stud. Pure Math., vol. 1, 1980, pp. 131-180.
[R3] Reid, M, *Young person's guide to canonical singularities*, Algebraic Geometry-Bowdoin 1985, vol. 46, 1987, pp. 333-414.
[U] Ueno, K, *Classification theory of algebraic varieties and compact complex spaces*, Lecture Notes in Mathematics **439**.
[Z] Zhang, Q, *Global holomorphic one-forms on projective manifolds with ample canonical bundles*, Journal of Algebraic Geometry **6** (1997), 777-787.

Department of Mathematics, University of Texas, Arlington, TX 76019
E-mail address: luo@uta.edu

Department of Mathematics, University of Missouri, Columbia, MO 65211
E-mail address: qi@math.missouri.edu

Contemporary Mathematics
Volume **386**, 2005

A NOTE ON TOROIDALIZATION: THE PROBLEM OF RESOLUTION OF SINGULARITIES OF MORPHISMS IN THE LOGARITHMIC CATEGORY

KENJI MATSUKI*

Contents

§0. Introduction

0.1. **Prologue**. The purpose of this note is to discuss the following problem of "toroidalization": Given a (proper) morphism $f : X \to Y$ between nonsingular varieties, can we find sequences of blowups with smooth centers $\pi_X : X' \to X$ and $\pi_Y : Y' \to Y$ such that the induced map $f' : X' \to Y'$ is a toroidal morphism (See §1 for the precise definition of a morphism to be toroidal.) ?

$$\begin{array}{ccc} X & \stackrel{\pi_X}{\longleftarrow} & X' \\ {\scriptstyle f}\downarrow & & \downarrow{\scriptstyle f'} \\ Y & \stackrel{\pi_Y}{\longleftarrow} & Y' \end{array}$$

The problem of toroidalization evolved along with the problem of semi-stable reduction and with the problem of factorization of birational maps, among others. The motivation behind is that, via toroidalization, we should be able to reduce the original problem to a combinatorial one, utilizing the correspondence between the geometry of toric varieties (toroidal embeddings) and the combinatorics of convex bodies. As such, it may seem as though the problem of toroidalization is a mere working hypothesis. The theme of this note is, however, to recognize its intrinsic importance, by presenting and reformulating it as the problem of resolution of singularities of morphisms in the logarithmic category. (See §1 for the precise meaning.) This view point was first communicated to the author by Prof. Dan Abramovich, though apparently it had been known to the experts for a while (cf. Kato[20]Abramovich-Karu[4]).

0.2. **A naive question**. A naive question can be posed in the following way: Given a (proper and surjective) morphism $f : X \to Y$ between nonsingular varieties,

*The author was partially supported by NSF grant DMS-0100598.

by choosing suitable "modifications" $\pi_X : X' \to X$ and $\pi_Y : Y' \to Y$, how "nice" can one make the induced morphism $f' : X' \to Y'$? (We require that we choose modifications π_X and π_Y in such a way that f' is a morphism, which a priori is just a rational map.) Of course, in order to make this question meaningful, one has to make explicit the mathematical specifications of the words "modifications" and "nice".

When $\dim Y = 1$ and we specify the morphism f' to be "nice" if every fiber is reduced with simple normal crossings, while the modification π_Y is restricted to a finite morphism (between nonsingular varieties) and the modification π_X is restricted to a sequence of blowups with smooth centers (possibly via normalization after base change), the question becomes the problem of semi-stable reduction (over a base of dimension one).

When $\dim Y = 1$ and we specify the morphism f' to be "nice" if every fiber has only simple normal crossings, while the modification π_Y is restricted to an isomorphism and the modification π_X is restricted to a sequence of blowups with smooth centers, the question becomes the problem of (embedded) resolution of hypersurface singularities.

When f is birational and we specify the morphism f' to be "nice" if it is an isomorphism, while the modifications π_X and π_Y are restricted to sequences of blowups with smooth centers, the question is the strong factorization conjecture of birational maps.

It should be warned, however, that the way we pose our question is very naive: for certain choices of the specifications of the word "modifications" or "nice", we often obtain a question to which the answer is *no* in general. This can be observed, for example, when we specify the morphism f' to be "nice" if it is flat, while the modifications π_X and π_Y are restricted to sequences of blowups with smooth centers. This flattening process by blowups with smooth centers does not exist in general, even for a generically finite morphism between nonsingular surfaces $f : X \to Y$, as is demonstrated by a famous example by Abhyankar[1] for the failure of simultaneous resolution of singularities. (The reader is also invited to see an easy toric example by T. Katsura in Oda[27].) It is, meanwhile, an interesting question whether the above flattening process by blowups with smooth centers exists or not, under the extra condition that the original f is a morphism with connected fibers. (The author does not know whether the answer is affirmative or negative.)[1]

The problem of toroidalization appears as one form of the questions in the above setting, where we specify f' to be "nice" if f' is toroidal, while the modifications π_X and π_Y are restricted to sequences of blowups with smooth centers.

It should be mentioned and emphasized that Abramovich-Karu[4] succeeds in

[1]Shortly after the first draft of the manuscript was posted on the archive, Dan Abramovich and Kalle Karu sent e-mails to the author presenting some examples of a morphism $f : X \to Y$ with connected fibers between nonsingular varieties of $\dim X = 3$ and $\dim Y = 2$ such that f can NOT be modified into a flat morphism f' by sequences of blowups with smooth centers. Therefore, the answer is still negative in general even with the extra assumption of the morphism having connected fibers. We remark, however, in the case where $\dim X = \dim Y$ that the assumption of the morphism having connected fibers is equivalent to the condition of the morphism being birational, and hence that the existence of the flattening process as above is nothing but the strong factorization conjecture of birational maps.

making the morphism f' toroidal, starting from a given proper morphism f, but by *not* restricting the modifications π_X and π_Y to be only sequences of blowups with smooth centers.

0.3. **Algorithm**. The reformulation of toroidalization as log resolution, though simple and straightforward as it is, naturally leads one to the following algorithmic approach (See §2 for details.):

Step 1: Characterize the locus where the morphism is not toroidal as the logarithmic ramification locus, which can then be computed using the logarithmic differential forms.

Step 2: Choose the center of blowup downstairs on Y, which should be contained in the image of the logarithmic ramification locus under f.

Step 3: Apply an algorithm for canonical principalization to the pull-back upstairs on X of the defining ideal of the center chosen in Step 2 downstairs on Y.

If the induced morphism is toroidal, then we stop. If not, we go back to Step 1 with the original morphism replaced by the induced one.

Termination: Show that the algorithm terminates after finitely many steps.

In fact, in the case where $\dim X = \dim Y = 2$, there is essentially no choice involved in the above algorithm and hence is uniquely determined. Thus the only remaining problem is to show termination. We accomplish this goal by providing a local analysis of the algorithm and then showing that the logarithmic ramification locus *decreases* (under certain order, with respect to which the descending chain condition is satisfied).

A discussion of the higher dimensional case will be published elsewhere.

0.4. **A possible approach to the factorization problem**. The problem of toroidalization was conceived as a possible approach to the (strong) factorization conjecture of birational maps (between nonsingular varieties into a sequence of blowups and blowdowns with smooth centers) in two steps:

Step I. Given a proper birational morphism $f : X \to Y$ (if you start with a birational map $f : X \dashrightarrow Y$, then use the elimination of indeterminacy by Hironaka to reach the stage where f can be assumed to be a morphism), apply toroidalization to make $f' : X' \to Y'$ toroidal.

Step II. Factor the toroidal birational morphism $f' : X' \to Y'$ into a sequence consisting of blowups with smooth centers immediately followed by blowdowns with smooth centers, by applying the toroidal version of a combinatorial algorithm for strong factorization of toric birational maps.

(Today the strong factorization conjecture remains open, while the weak factorization conjecture, which is strong enough for most of the applications, is a theorem achieving Steps I and II in a slightly weaker sense:

Step I′: Given a proper birational morphism $f : X \to Y$, apply the theory of birational cobordism by Włodarczyk[29] and torification by Abramovich-Karu-Matsuki-Włodarczyk[6] to decompose f as a composite of toroidal birational maps $f_i : X_i \dashrightarrow Y_i$ for $i = 0, ..., l$ (where $X = X_0, Y_i = X_{i+1}$ for $i = 0, ..., l-1$, and $Y_l = Y$).

Step II′: Apply the toroidal version of the algorithm for weak factorization by Morelli (cf. Abramovich-Matsuki-Rashid[5]Matsuki[23]Włodarczyk[28]) to each toroidal birational map $f_i : X_i \dashrightarrow Y_i$.

We fall short of realizing strong factorization, since in Step I′ the toroidal structures for f_i's are not compatible with each other in general, and since in Step II′ we know only of the algorithm for weak (but not strong) factorization even for toric birational maps (in dimension three or more) for the moment.)

0.5. **Work of S.D.Cutkosky**. Even at the time when, inspired by the work of A.J. de Jong, we could only daydream of the above approach to the problem of factorization, S.D. Cutkosky[10] made a remarkable announcement of a proof for *local* factorization of birational maps in dimension 3. (The *local* version of the problem of factorization replaces a birational map between nonsingular varieties X and Y with birational local rings dominated by a fixed valuation, and replaces blowups along smooth centers with monoidal transforms followed by localization subordinate to the valuation.)

In a sequence of papers [9][10][11], Cutkosky brilliantly establishes the monomialization, which is the *local* version of toroidalization, in arbitrary dimension, thus completing Step I in the *local* case.

Recently Karu[19] succeeded in providing an algorithm for *local* strong factorization of toric (toroidal) birational maps in arbitrary dimension, extending the earlier work of Christensen[9] in dimension 3.

Thus the *local* strong factorization of birational maps, known also as the Abhyankar conjecture, is now a theorem according to the general scheme of establishing Steps I and II.

Moreover, Cutkosky[12] also establishes the *global* version, i.e., the toroidalization when $\dim X = 3$ and $\dim Y = 2$.

0.6. **Contents of our paper**. The simple purpose of this paper is to give a proof of toroidalization in dimension 2, guided by the principle **toroidalization = log resolution**, which has been the main point of our theme from the conception of the problem (cf. Abramovich-Karu-Matsuki-Włodarczyk[6] Matsuki[24]). It is the belief (prejudice) of the author that only thorugh this reformulation the ultimate solution would be achieved by finding a much-sought-for inductional structure needed to prove the general case of toroidalization.

The contents of the paper are as follows.

In §1, we present our formulation of the problem of toroidalization as the problem of resolution of singularities of morphisms in the logarithmic category. The key is the study of the behavior of the logarithmic differential forms. In §2, we present an essentially unique algorithm for toroidalization in the case where $\dim X = \dim Y = 2$. In §3, we present a proof for toroidalization in dimension 2. Several proofs are known for toroidalization in dimension 2 by now (cf. Abramovich-Karu[4] Abramovich-Karu-Matsuki-Włodarczyk[6]Akbulut-King[7]Cutkosky-Piltant[13] Matsuki[24]). However, all the existing ones use the strong factorization theorem of a (proper) birational *morphism* (via Castelnuovo's contractibility criterion of a (-1)-curve) in the process of verification. This has a fatal defect: not only one cannot hope to extend the process to higher dimension where the factorization of a proper birational *morphism* into blowups with smooth centers fails to hold, but also it would bring a logical loop in our approach to solving the problem of factorization through toroidalization, even in dimension 2. They also use the properness assumption of f in an essential way, and as a result, their analysis is local on Y

but global on X in nature. In contrast, we provide a detailed but rather elementary analysis of the algorithm, which is completely local both on Y and on X. As a consequence, we show toroidalization in dimension 2 without the properness assumption of f. We do not use the strong factorization theorem in the process. Therefore, establishing Steps I and II of the general scheme, we give a new proof of the strong factorization of birational maps in diemsnion 2.

The way we show that the algorithm for toroidalization teminates is also guided by the principle, rather than setting up invariants in an adhoc way. We show that the logarithmic ramification locus *decreases* (under certain order, with respect to which the descending chain condition is satisfied) in the process.

0.7. **Assumption on the base field.** In this paper, we assume that the base field k is an algebraically closed field of characteristic zero. The assumption of k being algebraically closed is purely for the simplicity of the presentation: the "canonical" nature of our algorithm (and of the algorithm for principalization of ideals) implies that the process prescribed over $\overline{k}$ is equivariant under the action of the Galois group $\mathrm{Gal}(\overline{k}/k)$ even when k is not algebraically closed, and hence that it is actually defined over k (as long as the extension $\overline{k}/k$ is separable). Therefore, this assumption can be removed without much further ado. The assumption of k being of characteristic zero, however, is essential, because the problem of toroidalization as we formulate is too restrictive in positive characteristic (Some easy counter examples to this restrictive formulation can be found in positive characteristic. See, e.g. Cutkosky-Piltant[13].) and hence needs modification.

0.7. **Acknowledgements.** The author would like to thank Dan Abramovich for guidance and help in the subject. Thanks are also due to Kalle Karu and Jarosław Włodarczyk for many invaluable comments. The author would like to thank the organizers of the Barrett conference held at the University of Tennessee at Knoxville in April 2002, especially Yasuyuki Kachi, Shashikant Mulay and Pavlos Tzermias, for providing a mathematically stimulating atmosphere through the excellent lecture series by Yujiro Kawamata and Yasutaka Ihara.

§1. Formulation of toroidalization as resolution of singularities of morphisms in the logarithmic category

1.1. **Outline of this section.** In this section, we formulate the problem of toroidalization and the problem of resolution of singularities of morphisms in the logarithmic category, and show that a solution to the latter would imply a solution to the former. (It could be said that, modulo some minor technical points, the two problems are essentially equivalent.)

For this purpose, we define when a morphism $f : X \to Y$ between nonsingular varieties is toroidal, or in the presence of the prescribed structures as nonsingular toroidal embeddings, when a morphism $f : (U_X, X) \to (U_Y, Y)$ between nonsingular toroidal embeddings is toroidal (with respect to the given toroidal structures (U_X, X) and (U_Y, Y)). We also define, in the presence of the logarithmic structures as nonsingular toroidal embeddings, when a morphism $f : (U_X, X) \to (U_Y, Y)$ in the logarithmic category is (log) smooth. (In order to make a clear distinction between the notion of a morphism being "smooth" in the usual category and the notion of a morphism being "smooth" in the logarithmic category, we always use the word "log smooth" for the latter.) One of the main points of the section is, though it is quite straightforward, to show that the condition of f being toroidal is equivalent to f being log smooth.

1.2. **Basic definitions.**

1.2.1. *Logarithmic category.* An object in the logarithmic category (of nonsingular toroidal embeddings) is a nonsingular toroidal embedding (U_X, X), i.e., a pair consisting of a nonsingular variety X and a dense open subset U_X such that $D_X = X - U_X$ is a divisor with only simple normal crossings.

(A divisor D_X with only simple normal crossings consists of *smooth* irreducible components and satisfies the condition that at any closed point $p \in D_X$ there exists an open analytic neighborhood $p \in U_p \subset X$ with a system of regular parameters $(x_1, ..., x_d)$ of $\widehat{\mathcal{O}_{X,p}}$ such that

$$D_X \cap U_p = \{ \prod_{m \in M} x_m = 0\}$$

for some subset $M \subset \{1, ..., d = \dim X\}$.)

A morphism $f : (U_X, X) \to (U_Y, Y)$ in the logarithmic category is a morphism between nonsingular toroidal embeddings such that:

(a) $f : X \to Y$ is a dominant morphism between nonsingular varieties,
(b) $D_X = f^{-1}(D_Y)$,
(c) $f|_{U_X} : U_X \to U_Y$ is a smooth morphism.

1.2.2. *Definition of a morphism being toroidal.* Let $f : (U_X, X) \to (U_Y, Y)$ be a morphism in the logarithmic category of nonsingular toroidal embeddings. We say f is toroidal if for any $p \in X$ with $q = f(p) \in Y$ there exist a toric (equivariant) morphism $\varphi : (T_V, V) \to (T_W, W)$ between nonsingular toric varieties with $T_V \subset V$ and $T_W \subset W$ indicating the tori, and points $p_V \in V$ and $q_W = \varphi(p_V) \in W$, such that we have isomorphisms of the completions of the local rings

$$\sigma : \widehat{\mathcal{O}_{X,p}} \overset{\sim}{\to} \widehat{\mathcal{O}_{V,p_V}}$$
$$\tau : \widehat{\mathcal{O}_{Y,q}} \overset{\sim}{\to} \widehat{\mathcal{O}_{W,q_W}}$$

with a commutative diagram

$$\begin{array}{ccc} \widehat{\mathcal{O}_{X,p}} & \overset{\sim}{\underset{\sigma}{\to}} & \widehat{\mathcal{O}_{V,p_V}} \\ f^* \uparrow & & \varphi^* \uparrow \\ \widehat{\mathcal{O}_{Y,q}} & \overset{\sim}{\underset{\tau}{\to}} & \widehat{\mathcal{O}_{W,q_W}} \end{array}$$

which induces, via inclusions, another commutative diagram among the ideals defining the boundary divisors

$$\begin{array}{ccc} \widehat{\mathcal{I}_{D_X,p}} & \overset{\sim}{\to} & \widehat{\mathcal{I}_{D_V,p_V}} \\ f^* \uparrow & & \varphi^* \uparrow \\ \widehat{\mathcal{I}_{D_Y,q}} & \overset{\sim}{\to} & \widehat{\mathcal{I}_{D_W,q_W}}. \end{array}$$

We say that a morphism $f : X \to Y$ between nonsingular varieties is toroidal if we can introduce the structures (U_X, X) and (U_Y, Y) of nonsingular toroidal embeddings, i.e., we can find divisors $D_X = X - U_X \subset X$ and $D_Y = Y - U_Y \subset Y$ with only simple normal crossings, such that $f : (U_X, X) \to (U_Y, Y)$ is a toroidal morphism in the logarithmic category.

We also invite the reader to look at Abramovich-Karu[4] Kempf-Knudsen-Mumford-Saint-Donat[21] for reference.

1.2.3. *Definition of a morphism being log smooth.* Let $f : (U_X, X) \to (U_Y, Y)$ be a morphism in the logarithmic category of nonsingular toroidal embeddings. We say f is log smooth if the natural homomorphism

$$f^*\{\wedge^{\dim Y}\Omega_Y^1(\log D_Y)\} \wedge^{\dim X - \dim Y} \Omega_X^1(\log D_X)\} \to \wedge^{\dim X}\Omega_X^1(\log D_X)$$

is surjective.

1.3. Equivalence of "toroidal" and "log smooth".

Proposition 1.3.1. *Let $f : (U_X, X) \to (U_Y, Y)$ be a morphism in the logarithmic category of nonsingular toroidal embeddings. Then f is toroidal if and only if it is log smooth.*

Proof.

We only demonstrate a proof in the case where $\dim X = \dim Y = 2$. We make a further assumption that at the point $p \in D_X \subset X$ of our concern two irreducible components of D_X meet and also that at $q = f(p) \in D_Y \subset Y$ two irreducible components of D_Y meet. (In the terminology of §3, the points p and q are of type 2_p and 2_q.) Though simple, the consideration of this typical case captures the essence of the idea. (We refer the reader to Kato[20] for a general proof, though it is not so much more difficult to deal with the general case via local analysis than with the typical case.)

We choose a system of regular parameters (x_1, x_2) of $\widehat{\mathcal{O}_{X,p}}$ in an analytic neighborhood U_p of p (resp. (y_1, y_2) of $\widehat{\mathcal{O}_{Y,q}}$ in an analytic neighborhood U_q of q) such that it is compatible with the logarithmic structure, i.e., $D_X \cap U_p = \{x_1 x_2 = 0\}$ (resp. $D_Y \cap U_q = \{y_1 y_2 = 0\}$).

By condition (b) $D_X = f^{-1}(D_Y)$ imposed on a morphism in the logarithmic category, we have

$$\begin{cases} f^* y_1 = u \cdot x_1^a x_2^b \\ f^* y_2 = v \cdot x_1^c x_2^d \end{cases}$$

where $a, b, c, d \in \mathbb{Z}_{\geq 0}$ and $u, v \in \widehat{\mathcal{O}_{X,p}}^{\times}$ are units.

We compute

$$\begin{aligned} f^*(d\log y_1 \wedge d\log y_2) &= f^*\left(\frac{dy_1 \wedge dy_2}{y_1 \cdot y_2}\right) \\ &= d\log f^* y_1 \wedge d\log f^* y_2 \\ &= \left(\frac{du}{u} + a\frac{dx_1}{x_1} + b\frac{dx_2}{x_2}\right) \wedge \left(\frac{dv}{v} + c\frac{dx_1}{x_1} + d\frac{dx_2}{x_2}\right) \\ &= r_{\log} \cdot \frac{dx_1 \wedge dx_2}{x_1 \cdot x_2} \end{aligned}$$

where

$$r_{\log} = \det \begin{bmatrix} a & b \\ c & d \end{bmatrix} + (\text{higher terms i.e. } = 0 \text{ at p}).$$

Therefore, we conclude

$$\begin{aligned}
&f \text{ being log smooth} \\
&\iff R_{\log} := \operatorname{div}(r_{\log}) = 0 \\
&\iff \exists\ (x_1, x_2)\ \&\ (y_1, y_2) \text{ s.t.} \begin{cases} f^* y_1 = u \cdot x_1^a x_2^b \\ f^* y_2 = v \cdot x_1^c x_2^d \end{cases} \\
&\qquad \text{where } \det \begin{bmatrix} a & b \\ c & d \end{bmatrix} \neq 0 \text{ and } u, v \in \widehat{\mathcal{O}_{X,p}}^{\times} \text{ are units.} \\
&\iff \exists\ (x_1, x_2)\ \&\ (y_1, y_2) \text{ s.t.} \begin{cases} f^* y_1 = x_1^a x_2^b \\ f^* y_2 = x_1^c x_2^d \end{cases} \text{where } \det \begin{bmatrix} a & b \\ c & d \end{bmatrix} \neq 0 \\
&\iff f \text{ being toroidal.}
\end{aligned}$$

It is obvious that the third condition implies the second, while one can see that the second implies the third by replacing x_1 with $u^\alpha v^\gamma \cdot x_1$ and x_2 with $u^\beta v^\delta \cdot x_2$, where the equation

$$\begin{bmatrix} a & b \\ c & d \end{bmatrix} \begin{bmatrix} \alpha & \gamma \\ \beta & \delta \end{bmatrix} = \begin{bmatrix} 1 & 0 \\ 0 & 1 \end{bmatrix}$$

has a (unique) solution $\begin{bmatrix} \alpha & \beta \\ \gamma & \delta \end{bmatrix} \in M_2(\mathbb{Q})$ because $\det \begin{bmatrix} a & b \\ c & d \end{bmatrix} \neq 0$.

This completes the proof.

1.4. **Formulation of the main conjectures**.

1.4.1. *Toroidalization Conjecture.* Let $f : X \to Y$ be a dominant morphism between nonsingular varieties. Then there should exist sequences of blowups with smooth centers $\pi_X : X' \to X$ and $\pi_Y : Y' \to Y$ such that the induced map $f' : X' \to Y'$ is a toroidal morphism.

1.4.2. *Resolution of singularities of morphisms in the logarithmic category.* Let $f : (U_X, X) \to (U_Y, Y)$ be a morphism in the logarithmic category of nonsingular toroidal embeddings. Then there exist sequences of blowups with permissible centers $\pi_X : (U_{X'}, X') \to (U_X, X)$ and $\pi_Y : (U_{Y'}, Y') \to (U_Y, Y)$ such that the induced map $f' : (U_{X'}, X') \to (U'_Y, Y')$ is log smooth. (Note that a center is called permissible if it is smooth and locally defined by a part of a system regular parameters compatible with the log structure.)

1.5. **Conclusion**.

An affirmative solution to the problem of resolution of singularities of morphisms in the logarithmic category implies an affirmative solution to the toroidalization conjecture.

In fact, let $f : X \to Y$ be a dominant morphism between nonsingular varieties. If f is smooth, we can take $U_X = X$ and $U_Y = Y$ for the assertion of the toroidalization conjecture. If f is not smooth, we look at the natural homomorphism

$$f^*\{\wedge^{\dim Y}\Omega_Y^1\} \wedge^{\dim X - \dim Y} \Omega_X^1 \twoheadrightarrow \mathcal{I}_R \otimes \wedge^{\dim X}\Omega_X^1 \hookrightarrow \wedge^{\dim X}\Omega_X^1$$

where $\mathcal{I}_R$ is the ideal defining the ramification locus R. We take a sequence of blowups with smooth centers $\tau_Y : \tilde{Y} \to Y$ so that $D_{\tilde{Y}} := \tau_Y^{-1}(f(R))$ is a divisor with only simple normal crossings. Then we take a sequence of blowups with smooth centers $\sigma_X : \overline{X} \to X$ so that the induced rational map $\overline{f} : \overline{X} \to \tilde{Y}$ is a morphism. Finally, we take a sequence of blowups with smooth centers $\tau_X : \tilde{X} \to \overline{X}$ so that $D_{\tilde{X}} := \tau_X^{-1}\overline{f}^{-1} D_{\tilde{Y}}$ is a divisor with only simple normal crossings. We have now a morphism $\tilde{f} : (U_{\tilde{X}} := \tilde{X} \setminus D_{\tilde{X}}, \tilde{X}) \to (U_{\tilde{Y}} := \tilde{Y} \setminus D_{\tilde{Y}}, \tilde{Y})$ in the logarithmic category of nonsingular toroidal embeddings. Now it is clear that a solution for resolution of singularities of the morphism $\tilde{f}$ in the logarithmic category would imply a solution for toroidalization of the original morphism f by construction.

§2. Algorithm for toroidalization

2.1. **Algorithm for toroidalization in the case where** $\dim \mathbf{X} = \dim \mathbf{Y} = \mathbf{2}$. The purpose of this section is to describe an algorithm for toroidalization of a dominant morphism $f : X \to Y$ between nonsingular varieties in the case where $\dim X = \dim Y = 2$. We make a couple of remarks at the end of the section regarding possible algorithms in higher dimension.

2.1.1. *Setting.* As in the formulation of the problem of toroidalization in §1, let $f : (U_X, X) \to (U_Y, Y)$ be a morphism in the logarithmic category of nonsingular toroidal embeddings in the case where $\dim X = \dim Y = 2$, that is to say:
(a) $f : X \to Y$ is a dominant morphism between nonsingular varieties of $\dim X = \dim Y = 2$,
(b) $D_X = f^{-1}(D_Y)$, and
(c) the restriction of f to the open subset U_X, denoted by $f|_{U_X}$, is a smooth morphism.

2.1.2. *Description of the algorithm.* Our algorithm for toroidalization proceeds as follows:

Step 1

We write down the natural homomorphism among the logarithmic differential forms

$$f^*\{\wedge^{\dim Y}\Omega_Y^1(\log D_Y)\} \wedge^{\dim X - \dim Y} \Omega_X^1(\log D_X) \twoheadrightarrow \mathcal{I}_{R_{\log}} \otimes \wedge^{\dim X}\Omega_X^1(\log D_X)$$
$$\hookrightarrow \wedge^{\dim X}\Omega_X^1(\log D_X)$$

where $\mathcal{I}_{R_{\log}}$ is the ideal defining the logarithmic ramification locus.

Note that in our case where $\dim X = \dim Y = 2$, or more generally when $f : X \to Y$ is a generically finite morphism, we have the usual logarithmic ramification formula between the log canonical divisors

$$K_X + D_X = f^*(K_Y + D_Y) + R_{\log},$$

where $R_{\log}$ is the logarithmic ramification divisor (cf. Iitaka[17]). The natural homomorphism above corresponds to the inclusion of line bundles

$$0 \to \mathcal{O}_X(f^*(K_Y + D_Y)) = \mathcal{O}_X(-R_{\log} + K_X + D_X)$$
$$\hookrightarrow \mathcal{O}_X(K_X + D_X).$$

Step 2

We look at $f(R_{\log})$, the image of the logarithmic ramification locus.

The special feature, which is a consequence of the assumption $\dim \mathbf{Y} = \mathbf{2}$, is that $\mathbf{f}(\mathbf{R}_{\log})$ consists of finitely many points.

Blowup Y with center $f(R_{\log})$ (or more precisely speaking, blowup Y along the defining ideal $\mathcal{I}_{f(R_{\log})}$ of the image of the logarithmic ramification locus with the reduced structure).

Step 3

Apply an algorithm for canonical principalization (See Remark 2.2.2 for the precise meaning.) to the ideal $f^{-1}(\mathcal{I}_{f(R_{\log})}) \cdot \mathcal{O}_X$

$$\begin{array}{ccc} (U_X, X) = (U_{X_0}, X_0) & \overset{\text{canonical principalization}}{\underset{\mathrm{cp}_1}{\longleftarrow}} & (U_{X_1}, X_1) \\ {\scriptstyle f=f_0}\downarrow & & {\scriptstyle f_1}\downarrow \\ (U_Y, Y) = (U_{Y_0}, Y_0) & \overset{\text{Blowup } f(R_{\log})}{\underset{\mathrm{bp}_1}{\longleftarrow}} & (U_{Y_1}, Y_1) \end{array}$$

Note that by the universal property of blowup for principalization, the induced rational map f_1 is guaranteed to be a morphism.

If the induced morphism $f_1 : (U_{X_1}, X_1) \to (U_{Y_1}, Y_1)$ is toroidal, we end the algorithm.

If the induced morphism $f_1 : (U_{X_1}, X_1) \to (U_{Y_1}, Y_1)$ is not toroidal, then we go back to Step 1 of the algorithm with $f := f_1$ and repeat the process.

2.2. Remarks on the algorithm.

2.2.1. *A morphism is always toroidal over Y in codimension one.* It is straightforward to observe that, if $\dim Y = 1$, then $R_{\log} = \emptyset$ and hence also $f(R_{\log}) = \emptyset$.

In fact, let $p \in D_X \subset X$ be an arbitrary closed point with $q = f(p) \in D_Y \subset Y$ its image. Let $f^* : \widehat{\mathcal{O}_{Y,q}} \to \widehat{\mathcal{O}_{X,p}}$ be the induced homomorphism between the completions of the local rings. We take some systems of regular parameters $(x_1, ..., x_d)$ of $\widehat{\mathcal{O}_{X,p}}$ and (y_1) of $\widehat{\mathcal{O}_{Y,q}}$, compatible with the logarithmic structures. (That is to say, for some analytic neighborhoods U_p of p and U_q of q, we have

$$D_X \cap U_p = \{\prod_{m \in M} x_m = 0\}$$
$$D_Y \cap U_q = \{y_1 = 0\}$$

for a subset $M \subset \{1, ..., d\}$.)

We observe

$$\begin{aligned} f \text{ is toroidal} &\Longleftrightarrow \exists \ (x_1, ..., x_d) \ \& \ (y_1) \ \text{ s.t.} \\ & f^* y_1 = x_1^{a_1} \cdots x_d^{a_d} = \prod x_i^{a_i} \text{ with } a_i \geq 0 \ \forall i \quad \text{and} \quad a_{i_o} \neq 0 \text{ for some } i_o \\ &\Longleftrightarrow \exists \ (x_1, ..., x_d) \ \& \ (y_1) \ \text{ s.t.} \\ & f^* y_1 = u \cdot x_1^{a_1} ... x_d^{a_d} = u \cdot \prod x_i^{a_i} \text{ with } a_i \geq 0 \ \forall i \quad \text{and} \quad a_{i_o} \neq 0 \text{ for some } i_o, \\ & \text{where } u \text{ is a unit.} \end{aligned}$$

(It is obvious that the second condition implies the last, while one can see that the last condition implies the second by, e.g., replacing x_{i_o} with $u^{1/i_o} \cdot x_{i_o}$, where $u^{1/i_o} \in \widehat{\mathcal{O}_{X,p}}$ exists because of the assumption of the base field k being algebraically closed of characteristic zero.)

The last condition obviously holds by condition (b) $D_X = f^{-1}(D_Y)$ imposed on a morphism in the logarithmic category.

Therefore, f is toroidal.

This observation immediately implies that, for an arbitrary dominant morphism $f : (U_X, X) \to (U_Y, Y)$ in the logarithmic category, there is no locus of logarithmic ramification over a generic point of codimension one, i.e.,

$$\text{codim}_Y f(R_{\log}) \geq 2.$$

This is why $f(R_{\log})$ consists of finitely many points in the case where $\dim Y = 2$.

2.2.2. *Canonical principalization.* By an algorithm for canonical principalization, we mean a specific and fixed algorithm which, to a given ideal $\mathcal{I}$ on a nonsingular toroidal embedding (U_X, X) with support contained in the boundary divisor, i.e., $\text{Supp}\mathcal{O}_X/\mathcal{I} \subset D_X = X \setminus U_X$, assigns a uniquely determined sequence of blowups

$$\begin{aligned} (U_X, X) = (U_{X_0}, X_0) &\overset{\pi_1}{\leftarrow} (U_{X_1}, X_1) \leftarrow ... \\ ... \leftarrow (U_{X_{i-1}}, X_{i-1}) &\overset{\pi_i}{\leftarrow} (U_{X_i}, X_i) \leftarrow ... \\ ... &\overset{\pi_l}{\leftarrow} (U_{X_l}, X_l) \end{aligned}$$

with permissible centers $Y_{i-1} \subset D_{X_{i-1}} \subset X_{i-1}$ (The adjective "permissible" means, by definition, that Y_{i-1} is smooth and that for any closed point $p \in Y_{i-1} \subset D_{X_{i-1}} \subset X_{i-1}$ there exists an analytic neighborhood U_p with a system of regular parameters $(x_1, ..., x_d)$ of $\widehat{\mathcal{O}_{X_{i-1},p}}$ such that

$$\begin{aligned} D_{i-1} \cap U_p &= \{ \prod_{m \in M} x_m = 0\} \\ Y_{i-1} \cap U_p &= \cap_{m \in N} \{x_m = 0\} \end{aligned}$$

for some subsets $M, N \subset \{1, ..., d = \dim X\}$.)

We require as a result of principalization that the ideal $\pi^{-1}\mathcal{I} \cdot \mathcal{O}_{X_l}$ is principal (where $\pi = \pi_l \circ \cdots \circ \pi_1$), i.e., for any closed point $p \in X_l$ there exists an analytic neighborhood U_p with a system of regular parameters $(x_1, ..., x_d)$ of $\widehat{\mathcal{O}_{X_l,p}}$ such that

$$D_l \cap U_p = \{ \prod_{m \in M} x_m = 0\}$$
$$\pi^{-1}\mathcal{I} \cdot \mathcal{O}_{X_l}|_{U_p} = (\prod_{m \in L} x_m^{a_m})$$

for some subsets $M, L \subset \{1, ..., d = \dim X\}$.

The adjective "canonical" is used to add the following two more requirements to the algorithm:

1. (Stability under pull-back by smooth morphisms) If $f : (U_X, X) \to (U_Y, Y)$ is a morphism in the logarithmic category between nonsingular toroidal embeddings with $f : X \to Y$ being a smooth morphism (in the usual sense), then the process for principalization of the ideal $\mathcal{I}_X = f^*\mathcal{I}_Y \subset f^*\mathcal{O}_Y = \mathcal{O}_X$ specified by the algorithm on (U_X, X) should coincide with the pull-back (i.e., the one obtained by taking the Cartesian product "$\times_Y X$") of the process for principalization of the ideal $\mathcal{I}_Y$ specified by the algorithm on (U_Y, Y).

It should be noted that this requirement of the algorithm for principalization being stable under pull-back by smooth morphisms implies equivariance: if a group (finite or algebraic) acts on (U_X, X) and the ideal $\mathcal{I}$ is invariant under the action, then the process of canonical principalization is equivariant, i.e., the action lifts to the blowups in the process. This can be seen easily by applying this requirement to the two smooth morphisms, $pr : X \times G \to X$ the projectition and $\mu : X \times G \to X$ the multiplication map representing the action, and by observing that the process of canonical principalization on $X \times G$ coincides both with the pull-back by pr and with the pull-back by μ.

2. (Analytic nature of the algorithm) Suppose that (U_{X_α}, X_α) with $\mathcal{I}_\alpha$ and (U_{X_β}, X_β) with $\mathcal{I}_\beta$ are gievn as above. Suppose that we have closed points $p_\alpha \in X_\alpha$ and $p_\beta \in X_\beta$ with an isomorphism between the completions

$$\widehat{\mathcal{O}_{X_\alpha,p_\alpha}} = \mathcal{O}_{X_\alpha,p_\alpha} \otimes \widehat{\mathcal{O}_{X_1,p_1}} \xrightarrow{\sim} \mathcal{O}_{X_\beta,p_\beta} \otimes \widehat{\mathcal{O}_{X_\beta,p_\beta}} = \widehat{\mathcal{O}_{X_\beta,p_\beta}}$$

which induces isomorphisms via inclusions

$$\widehat{\mathcal{I}_\alpha} = \mathcal{I}_\alpha \otimes \widehat{\mathcal{O}_{X_\alpha,p_\alpha}} \xrightarrow{\sim} \mathcal{I}_\beta \otimes \widehat{\mathcal{O}_{X_\beta,p_\beta}} = \widehat{\mathcal{I}_\beta}$$
$$\widehat{\mathcal{I}_{D_\alpha}} = \mathcal{I}_{D_\alpha} \otimes \widehat{\mathcal{O}_{X_\alpha,p_\alpha}} \xrightarrow{\sim} \mathcal{I}_{D_\beta} \otimes \widehat{\mathcal{O}_{X_\beta,p_\beta}} = \widehat{\mathcal{I}_{D_\beta}}.$$

Then the isomorphism

$$(U_{X_\alpha}, X_\alpha) \times \operatorname{Spec} \widehat{\mathcal{O}_{X_\alpha,p_\alpha}} \xrightarrow{\sim} (U_{X_\beta}, X_\beta) \times \operatorname{Spec} \widehat{\mathcal{O}_{X_\beta,p_\beta}}$$

lifts to the isomorphisms between the blowups in the process of principalization for the ideal $\mathcal{I}_\alpha$ on (U_{X_α}, X_α), pulled back by Spec $\widehat{\mathcal{O}_{X_\alpha,p_\alpha}}$, and the blowups in the process of principalization for the ideal $\mathcal{I}_\beta$ on (U_{X_β}, X_β), pulled back by Spec $\widehat{\mathcal{O}_{X_\beta,p_\beta}}$.

3. (Avoidance of blowing up the points where the ideal is already principal) If the ideal $\mathcal{I}$ is already principal at a closed point $p \in X$, then the algorithm for canonical principalization leaves a neighborhood of p untouched.

An important consequence of requirements 1 and 2 is that if we apply the algorithm for canonical principalization to a toroidal ideal, then the blowups in the process of principalization are all toroidal as well as the ideals which are the total transforms of the original toroidal ideal.

In our special case where $\dim X = \dim Y = 2$, *there is a unique algorithm for canonical principalization, namely, at each stage on* X_{i-1} *we only blow up the points where the total transform* $\mathcal{I}_{i-1}$ *of the ideal is not* principal.

2.2.3 *Notation for the logarithmic ramification locus.* Given a morphism $f : (U_X, X) \to (U_Y, Y)$ in the logarithmic category, if we want to emphasize that the logarithmic ramification locus $R_{\log}$ is associated to the morphism f, we write $R_{\log,f}$ instead of $R_{\log}$.

2.3. **Further remarks on the algorithm**

2.3.1. *Case where* $\dim X > 2$ *and* $\dim Y = 2$ Our algorithm formally makes sense even in the case $\dim X > 2$ as long as $\dim Y = 2$, adopting any algorithm for canonical principalization, such as the one prescribed by Bierstone-Milman[8] or Encinas-Villamayor[14][15] among others, satisfying the above requirements. However, the author cannot show that the algorithm terminates even in the case $\dim X = 3$.

The major difference between the case where $\dim X > 2$ and the case where $\dim X = 2$ is that in the latter the defining ideal for the logarithmic ramification locus is always principal whereas in the former may not be. In fact, even when $\dim X > 2$, if the defining ideal for the logarithmic ramification locus happens to be principal, a combinatorial analysis very similar to the one in the case where $\dim X = 2$ would yield toroidalization. Therefore, one could try to formulate an algorithm where we first aim at principalizing the defining ideal for the logarithmic ramification locus. An interpretation of the monumental work by Cutkosky[12] could be given from this point of view. Deatails will be published elsewhere.

2.3.2. *Case where* $\dim Y > 2$ We are completely at loss in the case where $\dim Y > 2$, even in the case where $\dim X = \dim Y = 3$ and hence the defining ideal for the logarithmic ramification locus is principal. The stalemate is a result of our (or rather, the author's) lack of understanding of the expected inductional structure in the scheme of toroidalization, though the logarithmic category should be tailor-made for such an inductional scheme via adjunction.

§3. Local analysis in the case where $\dim X = \dim Y = 2$

3.1. **Purpose of this section**. The purpose of this section is to present a detailed local analysis of what happens in $\boxed{\text{Steps } 1, 2, 3}$ of the algorithm 2.1 for toroidalization in the case where $\dim X = \dim Y = 2$, under the same setting as in 2.1.1, i.e., $f : (U_X, X) \to (U_Y, Y)$ is a morphism in the logarithimic category of nonsingular toroidal embeddings in the case where $\dim X = \dim Y = 2$ so that

(a) $f : X \to Y$ is a dominant morphism between nonsingular varieties of

$\dim X = \dim Y = 2$,
(b) $D_X = f^{-1}(D_Y)$, and
(c) the restriction of f to the open subset U_X, denoted by $f|_{U_X}$, is a smooth morphism.

3.2. *Type and Case Descriptions.* Let $p \in D_X \subset X$ be a closed point and $q = f(p) \in D_Y \subset Y$ its image. In the following analysis, we always choose q to be in the image of the logarithmic ramification locus, i.e., $q \in f(R_{\log})$.

We make our analysis sorting out the types and cases, based upon the following criteria:

∘ The morphism f is toroidal or not at p.
We put the letter T when it is toroidal, and the letter N when not.
∘ Description of the boundary divisors at p and q and their behavior with respect to f.

We have the following type descriptions on the boundary divisors at p and q (We take some systems of regular parameters (x_1, x_2) of $\widehat{\mathcal{O}_{X,p}}$ and (y_1, y_2) of $\widehat{\mathcal{O}_{Y,q}}$, compatible with the logarithmic structures, in analytic neighborhoods U_p of p and U_q of q, respectively.):

2_p: The point p is the intersection point of two irreducible components $G_1 = \{x_1 = 0\}$ and $G_2 = \{x_2 = 0\}$ with $G_1 \cup G_2 = D_X \cap U_p$ for some analytic neighborhood U_p of p.
1_p: The point p belongs to only one irreducible component $G_1 = \{x_1 = 0\}$ with $G_1 = D_X \cap U_p$ for some analytic neighborhood U_p of p, while $G_2 = \{x_2 = 0\} \not\subset D_X \cap U_p$.
2_q: The point q is the intersection point of two irreducible components $H_1 = \{y_1 = 0\}$ and $H_2 = \{y_2 = 0\}$ with $H_1 \cup H_2 = D_Y \cap U_q$ for some analytic neighborhood U_q of q
1_q: The point q belongs to only one irreducible component $H_1 = \{y_1 = 0\}$ with $H_1 = D_Y \cap U_q$ for some analytic neighborhood U_q of q, while $H_2 = \{y_2 = 0\} \not\subset D_Y \cap U_q$.

We have the following ten subcases (some of which, namely, Subcase 2_p1_q0 and Subcase 1_p2_q0, will be doomed impossible), where the third number indicates how many irreducible components of the boundary divisor map onto q:

Subcase 2_p2_q0: Neither G_1 nor G_2 maps onto q
Subcase 2_p2_q1: Only one of G_1 or G_2 (say, G_1) maps onto q
Subcase 2_p2_q2: Both G_1 and G_2 maps onto q
Subcase 1_p2_q0: G_1 does not map onto q (and hence maps onto H_1 (or H_2))
Subcase 1_p2_q1: G_1 maps onto q
Subcase 1_p1_q0: G_1 does not map onto q (and hence G_1 maps onto H_1)
Subcase 1_p1_q1: G_1 maps onto q
Subcase 2_p1_q0: Neither G_1 nor G_2 maps onto q
Subcase 2_p1_q1: Only one of G_1 or G_2 (say, G_1) maps onto q
Subcase 2_p1_q2: Both G_1 and G_2 maps onto q

3.3. *Case by case analysis.* Now we carry out a detailed analysis, in each (sub)case, of what happens when we blowup q in $\boxed{\text{Step 2}}$ and when we take the canonical principalization of $f^{-1}\mathcal{I}_q \cdot \mathcal{O}_X = f^{-1}m_q \cdot \mathcal{O}_X$ in $\boxed{\text{Step 3}}$.

We provide:

A) Coordinate Expression
B) Descriptions of the canonical principalization cp_1 and the induced morphism f_1
C) Conclusion on the behavior of the logarithmic ramification divisor

It turns out that what weighs more, for the purpose of sorting out the subcases, is the type description of the point q than that of p, and we regroup the subcases accordingly.

Remark on notation: When we write Subcases $* \, 2_q \, * \, T$, we indicate that we are dealing with all the subcases of type 2_p2_q0, 2_p2_q1, 2_p2_q2, 1_p2_q0, 1_p2_q1, collectively, with the extra condition that f is toroidal at p.

When we write 1_p2_q0 without mentioning whether f is toroidal or not at p, we indicate that we are dealing with all the possibilities in the subcase without a priori putting the extra assumption on whether f is toroidal or not.

3.3.1) Subcases $* \, 2_q \, *$: These consist of the following subcases:

Subcase $2_p 2_q 0$
Subcase $2_p 2_q 1$
Subcase $2_p 2_q 2$
Subcase $1_p 2_q 1$

Subcase $1_p 2_q 0$: This subcase does not happen. In fact, say G_1 maps onto H_1 (and hence does not map onto H_2). Note that in a neighborhood U_p of p we have $f^{-1}(H_2) \cap U_p \subset f^{-1}(D_Y) \cap U_p = D_X \cap U_p = G_1$. Now since $f^{-1}(H_2)$ is of pure codimension one, being the pull-back of a Cartier divisor, and since $p \in f^{-1}(H_2)$, we conclude $G_1 \cap U_p = f^{-1}(H_2) \cap U_p$. But this would imply either G_1 maps onto q or maps onto H_2, a contradiction!

3.3.1.A) Coordinate Expression

By the computation presented in the proof of Proposition 1.3.1, we conclude the following.

Subcases $2_p 2_q \, * T$: There exist systems of regular parameters (x_1, x_2) of $\widehat{\mathcal{O}_{X,p}}$ and (y_1, y_2) of $\widehat{\mathcal{O}_{Y,q}}$ compatible with the logarithmic structures such that

$$\begin{cases} f^* y_1 = x_1^a x_2^b \\ f^* y_2 = x_1^c x_2^d \end{cases} \quad \text{where} \quad \det \begin{bmatrix} a & b \\ c & d \end{bmatrix} \neq 0.$$

Subcases $2_p 2_q \, * N$: There exist systems of regular parameters (x_1, x_2) of $\widehat{\mathcal{O}_{X,p}}$ and (y_1, y_2) of $\widehat{\mathcal{O}_{Y,q}}$ compatible with the logarithmic structures such that

$$\begin{cases} f^* y_1 = u \cdot x_1^a x_2^b \\ f^* y_2 = v \cdot x_1^c x_2^d \end{cases} \quad \text{where} \quad \det \begin{bmatrix} a & b \\ c & d \end{bmatrix} = 0 \quad \text{with units} \quad u, v \in \widehat{\mathcal{O}_{X,p}}^{\times}.$$

Subcases $1_p 2_q \, *$: There exist systems of regular parameters (x_1, x_2) of $\widehat{\mathcal{O}_{X,p}}$ and (y_1, y_2) of $\widehat{\mathcal{O}_{Y,q}}$ compatible with the logarithmic structures, as described in 3.2, such that

$$\begin{cases} f^* y_1 = u \cdot x_1^a \\ f^* y_2 = v \cdot x_1^c \end{cases} \quad \text{where } a, c \in \mathbb{Z}_{>0} \text{ with units} \quad u, v \in \widehat{\mathcal{O}_{X,p}}^{\times}.$$

3.3.1.B) Descriptions of the canonical principalization cp_1 and the induced morphism f_1

Subcases $2_p 2_q \, * T$: We observe that the ideal $\mathcal{I}_q = m_q = (y_1, y_2)$ is toroidal (in a neighborhood of q) in these subcases and that the ideal $f^{-1}(y_1, y_2) \cdot \mathcal{O}_{X,p} = (x_1^a x_2^b, x_1^c x_2^d)$ is also toroidal (in a neighborhood of p). Therefore, by remark 2.2.2, the blowup $\mathrm{bp}_1 : (U_{Y_1}, Y_1) \to (U_{Y_0}, Y_0)$ is toroidal over a neighborhood of q, and so is the canonical principalization $\mathrm{cp}_1 : (U_{X_1}, X_1) \to (U_{X_0}, X_0) = (U_X, X)$ over a neighborhood of p. Since a composite of toroidal morphisms is again toroidal and so is a composite of a toroidal morphism with the inverse of a toroidal birational morphism (assuming that the composite is again a morphism, which is the case here), we conclude that $f_1 = \mathrm{bp}_1^{-1} \circ f_0 \circ \mathrm{cp}_1$ stays toroidal on the locus of (U_{X_1}, X_1) which sits over a neighborhood of p.

Subcases $2_p 2_q * N$: In these subcases, as we observed, we have

$$\det \begin{bmatrix} a & b \\ c & d \end{bmatrix} = 0,$$

that is to say, the vectors $(a,b), (c,d) \in \mathbb{Z}_{\geq 0} \times \mathbb{Z}_{\geq 0}$ are linearly dependent. (We may choose (y_1, y_2), without loss of generality, so that $a \geq c, b \geq d$. Note that $(a,b) \neq (0,0)$ and $(c,d) \neq (0,0)$.) This implies that the ideal $f^{-1}(y_1, y_2) \cdot \mathcal{O}_{X,p} = (x_1^c x_2^d)$ is already principal in a neighborhood of p. Therefore, by remark 2.2.2 the canonical principalization cp_1 leaves a neighborhood of p untouched. Also, by remark 2.2.2, in a neighborhood of p the map $f_1 = \mathrm{bp}_1^{-1} \circ f_0 \circ \mathrm{cp}_1 = \mathrm{bp}_1^{-1} \circ f_0$ is a morphism.

Subcases $1_p 2_q *$: In these subcases, the ideal $f^{-1}(y_1, y_2) \cdot \mathcal{O}_{X,p} = (x_1^a)$ (We may choose (y_1, y_2) without loss of generality so that $a \leq c$ in the coordinate expression.) is already principal in a neighborhood of p. Therefore, by remark 2.2.2 the canonical principalization cp_1 leaves a neighborhood of p untouched. Also, by remark 2.2.2, in a neighborhood of p the map $f_1 = \mathrm{bp}_1^{-1} \circ f_0 \circ \mathrm{cp}_1 = \mathrm{bp}_1^{-1} \circ f_0$ is a morphism.

3.3.1.C) Conclusion on the behavior of the logarithmic ramification divisor

In each of the subcases above, we conclude that over a neighborhood of the point q, since bp_1 is toroidal,

$$K_{Y_1} + D_{Y_1} = \mathrm{bp}_1^*(K_{Y_0} + D_{Y_0}) \quad \text{i.e.,} \quad \wedge^{\dim Y_1} \Omega^1_{Y_1}(\log D_{Y_1}) = \mathrm{bp}_1^* \wedge^{\dim Y_0} \Omega^1_{Y_0}(\log D_{Y_0}),$$

and hence that the logarithmic ramification divisor stays intact, i.e.,

$$R_{\log, f_0} = R_{\log, f_1}.$$

3.3.2) Subcase $1_p 1_q 0$: G_1 does not map onto q (and hence G_1 maps onto H_1)

3.3.2.A) Coordinate Expression

In this subcase, we show that f is necessarily toroidal at p. (Although this is essentially due to Abhyankar's lemma, we will not use it explicitly). We will show by some elementary calculation that there exist systems of regular parameters (x_1, x_2) of $\widehat{\mathcal{O}_{X,p}}$ and (y_1, y_2) on $\widehat{\mathcal{O}_{Y,q}}$, compatible with the logarithmic structures, such that

$$\begin{cases} f^* y_1 = x_1^a \\ f^* y_2 = x_2 \end{cases} \quad \text{with} \quad a > 0$$

using condition (c) $f|_{U_X}$ being smooth, imposed on a morphism in the logarithmic category (cf. 1.2.1).

First, we start with some systems of regular parameters as described in 3.2. Then since $f^{-1}(D_Y) = D_X$ where $D_Y \cap U_q = H_1 = \{y_1 = 0\}$ and $D_X \cap U_p = G_1 = \{x_1 = 0\}$, we have

$$f^* y_1 = u \cdot x_1^a \quad \text{for some unit } u \in \widehat{\mathcal{O}_{X,p}}^{\times} \quad \text{with} \quad a > 0.$$

By replacing x_1 with $u^{1/a} \cdot x_1$, we may assume

$$f^* y_1 = x_1^a.$$

Set

$$f^* y_2 = \Sigma_{i>0, j\geq 0, \det\begin{bmatrix} a & 0 \\ i & j \end{bmatrix} = 0} \alpha_{ij} x_1^i x_2^j + \Sigma_{i\geq 0, j>0, \det\begin{bmatrix} a & 0 \\ i & j \end{bmatrix} \neq 0} \alpha_{ij} x_1^i x_2^j,$$

where there exists $j > 0$ with $\alpha_{0j} \neq 0$, since x_1 should not divide $f^* y_2$ as $G_1 = \{x_1 = 0\}$ does not map onto q by the subcase assumption.

We set

$$j_o := \min\{j; \alpha_{0j} \neq 0\}.$$

The basic point of computing the ramification is the simple observation that

$$d(x_1^a x_2^b) \wedge d(x_1^c x_2^d) = \det \begin{bmatrix} a & b \\ c & d \end{bmatrix} \cdot x_1^{a+c-1} x_2^{b+d-1} dx_1 \wedge dx_2.$$

We compute

$$\begin{aligned} f^*(dy_1 \wedge dy_2) &= df^* y_1 \wedge df^* y_2 \\ &= d(x_1^a) \wedge d\left(\Sigma_{i>0, j\geq 0, \det\begin{bmatrix} a & 0 \\ i & j \end{bmatrix} = 0} \alpha_{ij} x_1^i x_2^j + \Sigma_{i\geq 0, j>0, \det\begin{bmatrix} a & 0 \\ i & j \end{bmatrix} \neq 0} \alpha_{ij} x_1^i x_2^j \right) \\ &= d(x_1^a) \wedge d\left(\Sigma_{i\geq 0, j>0, \det\begin{bmatrix} a & 0 \\ i & j \end{bmatrix} \neq 0} \alpha_{ij} x_1^i x_2^j \right) \\ &= (\Sigma_{i\geq 0, j>0, \det\begin{bmatrix} a & 0 \\ i & j \end{bmatrix} \neq 0} \det \begin{bmatrix} a & 0 \\ i & j \end{bmatrix} \cdot \alpha_{ij} x_1^{a+i-1} x_2^{j-1}) \cdot dx_1 \wedge dx_2 \\ &= x_1^{a-1} \cdot \left(\Sigma_{i\geq 0, j>0, \det\begin{bmatrix} a & 0 \\ i & j \end{bmatrix} \neq 0} \det \begin{bmatrix} a & 0 \\ i & j \end{bmatrix} \cdot \alpha_{ij} x_1^i x_2^{j-1} \right) \cdot dx_1 \wedge dx_2. \end{aligned}$$

Therefore, if $j_o > 1$, then f ramifies along

$$\{\left(\sum_{i\geq 0, j>0, \det\begin{bmatrix} a & 0 \\ i & j\end{bmatrix}\neq 0} \det\begin{bmatrix} a & 0 \\ i & j\end{bmatrix} \cdot \alpha_{ij} x_1^i x_2^{j-1}\right) = 0\}$$

other than along $\{x_1 = 0\}$, a contradiction to condition $f|_{U_X}$ being smooth! Therefore, we conclude

$$j_o = 1.$$

By replacing x_2 with $f^* y_2$, we obtain the desired systems of regular parameters.

Now since neither the divisor $\{y_2 = 0\}$ nor $\{x_2 = 0\}$ belongs to the boundary divisor defining the logarithmic structure, it is clear f is toroidal at p from the coordinate expression.

3.3.2.B) Descriptions of the canonical principalization cp_1 and the induced morphism f_1

We consider the 1st blowup $\mathrm{cp}_{1,1}$ of the canonical principalization cp_1, which is a sequence $\mathrm{cp}_1 = \mathrm{cp}_{1,1} \circ \cdots \circ \mathrm{cp}_{1,l}$ of blowups specified by the canonical principalization algorithm.

Diagram 3.3.2.B.1.

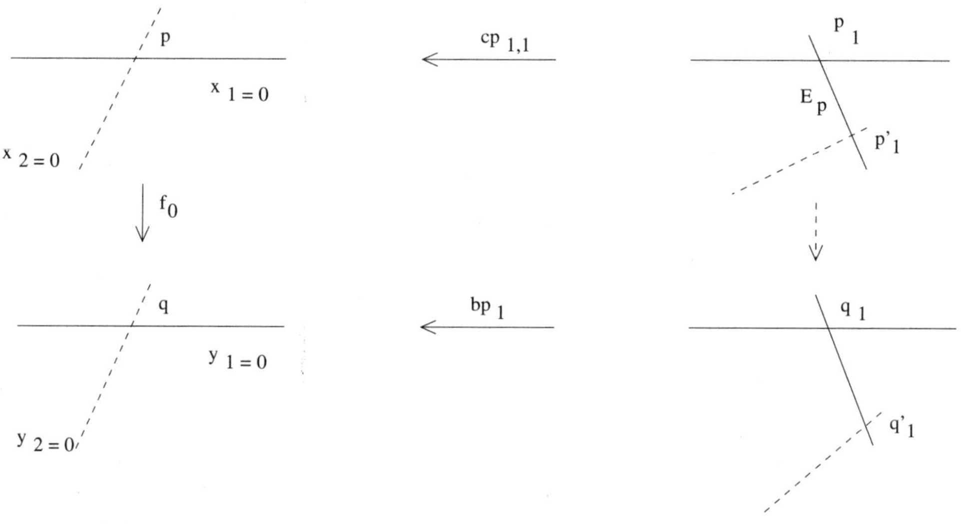

Claim 3.3.2.B.2.

a) The rational map $f_{1,1} = \mathrm{bp}^{-1} \circ f_0 \circ \mathrm{cp}_{1,1}$ is well-defined (regular), that is to say, the ideal $(f_0 \circ \mathrm{cp}_{1,1})^{-1}(y_1, y_2) \cdot \mathcal{O}_{X_{1,1}}$ is principal, except possibly at p'_1.

b) At p'_1, the morphism $f \circ \mathrm{cp}_{1,1}$ is in Subcase $1_{p'_1} 1_q 1$.

Proof.

a) Observe that the morphisms f_0 and $\mathrm{cp}_{1,1}$ and the ideal (y_1, y_2) are toroidal, with respect to the modified logarithmic structures obtained by adding $\{y_2 = 0\}$

and $\{x_2 = 0\}$ to the original boundary divisors. From this it follows easily that $f_{1,1}$ is regular, except possibly at p_1 and/or p_1'.

Now at p_1, we have a system of regular parameters $(\frac{x_1}{x_2}, x_2)$ with coordinate expression

$$\begin{cases} (f_0 \circ \mathrm{cp}_{1,1})^* y_1 = ((\frac{x_1}{x_2})x_2)^a \\ (f_0 \circ \mathrm{cp}_{1,1})^* y_2 = x_2, \end{cases}$$

which immediately implies that the ideal $(f_0 \circ \mathrm{cp}_{1,1})^{-1}(y_1, y_2) \cdot \widehat{\mathcal{O}_{X_{1,1},p_1}} = (x_2)$ is principal at p_1 and hence that $f_{1,1}$ is regular at p_1.

b) The verification for statement b) is immediate.

3.3.2.C) Conclusion on the behavior of the logarithmic ramification divisor

Claim 3.3.2.C.1.

a) The coefficient of G_1 in the logarithmic ramification divisor remains 0, i.e.,

$$\nu_{G_1}(R_{\log,f_0}) = \nu_{G_1}(R_{\log,f_1}) = 0.$$

b) The strict transform of the exceptional divisor E_p for $\mathrm{cp}_{1,1}$, *obtained by blowing up p, does not appear in $R_{\log,f_1}$, i.e.,*

$$\nu_{E_p}(R_{\log,f_1}) = 0.$$

More generally, none of the irreducible components E of the exceptional divisor for cp_1 *(i.e., none of the strict transforms of the exceptional divisors for* $\mathrm{cp}_{1,1}, \ldots, \mathrm{cp}_{1,l}$*) appear in $R_{\log,f_1}$, i.e.,*

$$\nu_E(R_{\log,f_1}) = 0.$$

Proof.

a) The verification for statement a) is obvious.

b) Take the standard neighborhood of q_1 with a system of regular parameters $(\frac{y_1}{y_2}, y_2)$. Then observing

$$\begin{cases} \nu_{E_p}(\frac{y_1}{y_2}) = a - 1 \geq 0 \\ \nu_{E_p}(y_2) = 1 > 0, \end{cases}$$

we conclude that the generic point of E_p maps into the standard neighborhood of q_1.

Observe that the morphisms $f_0, \mathrm{bp}_1, \mathrm{cp}_{1,1}$ are all toroidal with respect to the modified logarithmic structures obtained by adding $\{y_2 = 0\}$ and $\{x_2 = 0\}$ (and their pull-backs) to the original boundary divisors. The original logarithmic structures coincide with the modified ones in a neighborhood of the generic point of E_p and in the standard neighborhood of q_1. Therefore, $f_{1,1}$ is toroidal in a neighborhood of the generic point of E_p with respect to the original logarithmic structures, and hence

$$\nu_{E_p}(R_{\log,f_1}) = 0.$$

This proves the first part of statement b).

The second part of statement b) follows from Claim 3.3.2.B.2 b) and Claim 3.3.3.C.1 b).

3.3.3) Subcase $1_p 1_q 1$: G_1 maps onto q

3.3.3.A) Coordinate Expression

First, we start with some systems of regular parameters as described in 3.2. Then since $f^{-1}(D_Y) = D_X$ where $D_Y \cap U_q = H_1 = \{y_1 = 0\}$ and $D_X \cap U_p = G_1 = \{x_1 = 0\}$, we have

$$f^* y_1 = u \cdot x_1^a \quad \text{for some unit } u \in \widehat{\mathcal{O}_{X,p}}^{\times} \quad \text{with} \quad a > 0.$$

By replacing x_1 with $u^{1/a} \cdot x_1$, we may assume

$$f^* y_1 = x_1^a.$$

Set

$$f^* y_2 = \Sigma_{i>0, j\geq 0, \det \begin{bmatrix} a & 0 \\ i & j \end{bmatrix} = 0} \alpha_{ij} x_1^i x_2^j + \Sigma_{i\geq 0, j>0, \det \begin{bmatrix} a & 0 \\ i & j \end{bmatrix} \neq 0} \alpha_{ij} x_1^i x_2^j,$$

where no term of the form $x_2^j = x_1^0 x_2^j \quad (i = 0, j > 0)$ appears, since x_1 has to divide $f^* y_2$.

Now since $f|_{U_X}$ must be smooth by condition (c) imposed on a morphism in the logarithmic category (cf. 1.2.1) and by computing the ramification $f^*(dy_1 \wedge dy_2)/dx_1 \wedge dx_2$ as in 3.3.2.A, we conclude that $f^* y_2$ must be of the form

$$\begin{aligned} f^* y_2 &= \Sigma_{i \geq i_s > 0, j=0} \alpha_{i0} x_1^i \\ &\quad + \alpha_{i_o,1} x_1^{i_o} x_2^1 + \Sigma_{i \geq i_o > 0, j>0, (i,j) \neq (i_o,1)} \alpha_{ij} x_1^i x_2^j \end{aligned} \quad \text{with} \quad \alpha_{i_o,1} \neq 0,$$

where

$$i_o = \min\{i; \alpha_{ij} \neq 0, \det \begin{bmatrix} a & 0 \\ i & j \end{bmatrix} \neq 0\}$$

$$i_s = \min\{i; \alpha_{ij} \neq 0, \det \begin{bmatrix} a & 0 \\ i & j \end{bmatrix} = 0\}.$$

3.3.3.B) Descriptions of the canonical principalization cp_1 and the induced morphism f_1

Case $i_s \leq i_o$: In this case, we have

$$\begin{cases} f^* y_1 = x_1^a \\ f^* y_2 = v \cdot x_1^{i_s} \end{cases} \quad \text{with} \quad v \in \widehat{\mathcal{O}_{X,p}}^{\times} \quad \text{unit},$$

which implies $f^{-1}(y_1, y_2) \cdot \widehat{\mathcal{O}_{X,p}} = (x_1^a, x_1^{i_s}) = (x_1^{\min\{a, i_s\}})$ is principal. Therefore, cp_1 is an isomorphism in a neighborhood of p.

Case $i_s > i_o$: In this case, we have

$$\begin{cases} f^* y_1 = x_1^a \\ f^* y_2 = x_1^{i_o} \cdot (\Sigma_{i \geq i_s > 0, j=0} \alpha_{i0} x_1^{i - i_o} + \alpha_{i_o,1} x_2 + \Sigma_{i \geq i_o, j>0, (i,j) \neq (i_o,1)} \alpha_{ij} x_1^{i - i_o} x_2^j). \end{cases}$$

Replacing x_2 by $(\Sigma_{i \geq i_s > 0, j=0} \alpha_{i0} x_1^{i-i_o} + \alpha_{i_o,1} x_2 + \Sigma_{i \geq i_o, j>0, (i,j) \neq (i_o,1)} \alpha_{ij} x_1^{i-i_o} x_2^j)$, we have

$$\begin{cases} f^* y_1 = x_1^a \\ f^* y_2 = x_1^c x_2 \end{cases} \quad \text{with} \quad c = i_o.$$

Subcase $a < c+1$, i.e., $a \leq c$: In this subcase, the ideal $f^{-1}(y_1, y_2) \cdot \widehat{\mathcal{O}_{X,p}} = (x_1^a)$ is principal. Therefore, cp_1 is an isomorphism in a neighborhood of p.

Subcase $a \geq c+1$: We consider the 1st blowup $\mathrm{cp}_{1,1}$ of the canonical principalization cp_1, which is a sequence of blowups $\mathrm{cp}_1 = \mathrm{cp}_{1,1} \circ \cdots \circ \mathrm{cp}_{1,l}$ specified by the canonical principalization algorithm.

Diagram 3.3.3.B.1.

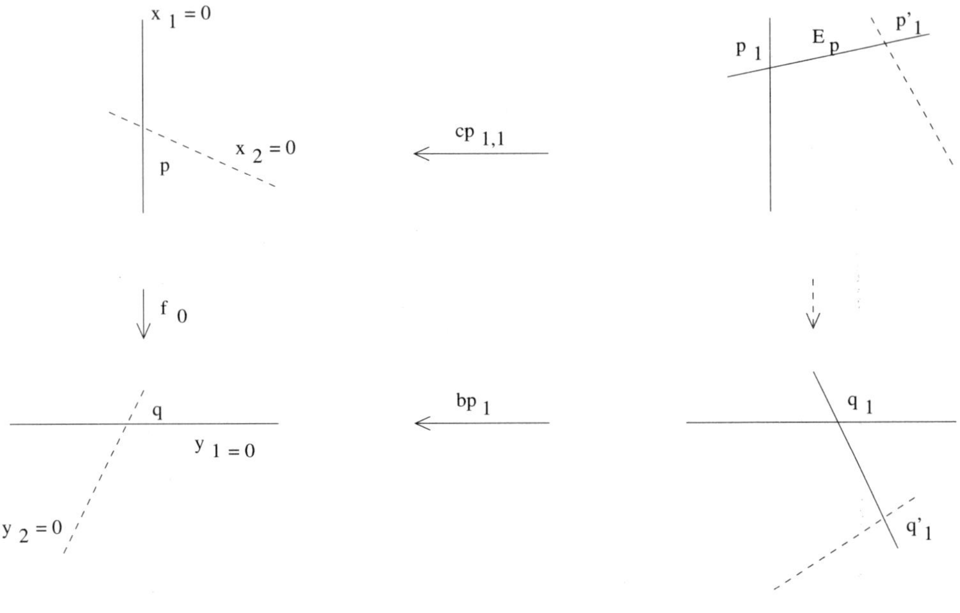

Claim 3.3.3.B.2. *The canonical principalization* cp_1 *is an isomorphism in a neighborhood of* p*, except under the subcase* $a \geq c+1$ *in the case* $i_o > i_1$*. In the exceptional subcase, we claim:*

a) The rational map $f_{1,1} = \mathrm{bp}^{-1} \circ f_0 \circ \mathrm{cp}_{1,1}$ *is well-defined (regular), that is to say, the ideal* $(f_0 \circ \mathrm{cp}_{1,1})^{-1}(y_1, y_2) \cdot \mathcal{O}_{X_{1,1}}$ *is principal, except possibly at* p_1'.

b) At p_1'*, the morphism* $f \circ \mathrm{cp}_{1,1}$ *is in Subcase* $1_{p_1'} 1_q 1$.

Proof.

The first part of the claim is already verified. So we have only to prove the assertions under the subcase $a \geq c+1$ in the case $i_o > i_1$.

a) Observe that the morphisms f_0 and $\mathrm{cp}_{1,1}$ and the ideal (y_1, y_2) are toroidal, with respect to the modified logarithmic structures obtained by adding $\{y_2 = 0\}$ and $\{x_2 = 0\}$ to the original boundary divisors. From this it follows easily that $f_{1,1}$ is regular, except possibly at p_1 and/or p_1'.

Now at p_1, we have a system of regular parameters $(\frac{x_1}{x_2}, x_2)$ with coordinate expression

$$\begin{cases} (f_0 \circ \mathrm{cp}_{1,1})^* y_1 = ((\dfrac{x_1}{x_2})x_2)^a = (\dfrac{x_1}{x_2})^a x_2^a \\ (f_0 \circ \mathrm{cp}_{1,1})^* y_2 = ((\dfrac{x_1}{x_2})x_2)^c x_2 = (\dfrac{x_1}{x_2})^c x_2^{c+1} \end{cases}$$

which immediately implies that the ideal $(f_0 \circ \mathrm{cp}_{1,1})^{-1}(y_1, y_2) = ((\frac{x_1}{x_2})^c x_2^{c+1})$ is principal at p_1 and hence that $f_{1,1}$ is regular at p_1.

b) The verification for statement b) is immediate.

3.3.3.C) Conclusion on the behavior of the logarithmic ramification divisor

Claim 3.3.3.C.1. *a) The coeffcient of G_1 in the logarithmic ramification divisor strictly drops. More precisely,*

$$\nu_{G_1}(R_{\log,f_1}) = \left. \begin{cases} i_o - \min\{i_o, i_s\} & \text{if } a \geq \min\{i_o, i_s\} \\ i_o - a & \text{if } a < \min\{i_o, i_s\} \end{cases} \right\} < i_o = \nu_{G_1}(R_{\log,f_0}).$$

b) The canonical principalization cp_1 is an isomorphism in a neighborhood of p, except under the subcase $a \geq c+1$ in the case $i_s > i_o$. In the exceptional subcase, the strict transform of the exceptional divisor E_p for $\mathrm{cp}_{1,1}$, obtained by blowing up p, does not appear in $R_{\log,f_1}$, that is to say, the coefficient of the strict transform of the exceptional divisor E_p is 0, i.e.,

$$\nu_{E_p}(R_{\log,f_1}) = 0.$$

More generally, none of the irreducible components E of the exceptional divisor for cp_1 (i.e., none of the strict transforms of the exceptional divisors for $\mathrm{cp}_1, \ldots, \mathrm{cp}_l$), appear in $R_{\log,f_1}$, i.e.,

$$\nu_E(R_{\log,f_1}) = 0.$$

Proof.

a) Firstly we compute the coefficient $\nu_{G_1}(R_{\log,f_0})$ of G_1 in the logarithmic ramification divisor $R_{\log,f_0}$.

$$\begin{aligned} \nu_{G_1}(R_{\log,f_0}) &= \nu_{G_1}\left(\frac{dy_1}{y_1} \wedge dy_2 / \frac{dx_1}{x_1} \wedge \frac{dx_2}{x_2}\right) \\ &= \nu_{G_1}\left(\frac{d(x_1^a)}{x_1^a} \wedge d\left(\Sigma_{i \geq i_o > 0, j > 0} \alpha_{ij} x_1^i x_2^j\right) / \frac{dx_1}{x_1} \wedge \frac{dx_2}{x_2}\right) \\ &= \nu_{G_1}\left(\frac{d(x_1^a)}{x_1^a} \wedge d(x_1^{i_o} x_2) / \frac{dx_1}{x_1} \wedge \frac{dx_2}{x_2}\right) = i_o. \end{aligned}$$

Secondly we compute the coefficient of (the strict transform of) G_1 in $R_{\log,f_1}$.

Subcase $a \geq \min\{i_o, i_s\}$: In this subcase, we have

$$\begin{cases} \nu_{G_1}(\dfrac{y_1}{y_2}) = a - \min\{i_o, i_s\} \geq 0 \\ \nu_{G_1}(y_2) = \min\{i_o, i_s\} > 0, \end{cases}$$

which implies that the generic point of (the strict transform of) G_1 maps under f_1 into the standard neighborhood of q_1 with a system of regular parameters $(\frac{y_1}{y_2}, y_2)$. Therefore, we compute

$$\begin{aligned}
\nu_{G_1}(R_{\log,f_1}) &= \nu_{G_1}\left(\frac{d(\frac{y_1}{y_2})}{(\frac{y_1}{y_2})} \wedge \frac{dy_2}{y_2} / \frac{d(\frac{x_1}{x_2})}{(\frac{x_1}{x_2})} \wedge \frac{dx_2}{x_2}\right) \\
&= \nu_{G_1}\left(\frac{dy_1}{y_1} \wedge dy_2 / \frac{dx_1}{x_1} \wedge \frac{dx_2}{x_2}\right) - \nu_{G_1}(y_2) \\
&= i_o - \min\{i_o, i_s\} < i_o.
\end{aligned}$$

Subcase $a < \min\{i_o, i_s\}$: In this subcase, we have

$$\begin{cases} \nu_{G_1}(y_1) = a > 0 \\ \nu_{G_1}(\frac{y_2}{y_1}) = \min\{i_o, i_s\} - a > 0, \end{cases}$$

which implies that the generic point of (the strict transform of) G_1 maps under f_1 into the standard neighborhood of q_1' with a system of regular parameters $(y_1, \frac{y_2}{y_1})$. Therefore, we compute

$$\begin{aligned}
\nu_{G_1}(R_{\log,f_1}) &= \nu_{G_1}\left(\frac{dy_1}{y_1} \wedge d(\frac{y_2}{y_1}) / \frac{d(\frac{x_1}{x_2})}{(\frac{x_1}{x_2})} \wedge \frac{dx_2}{x_2}\right) \\
&= \nu_{G_1}\left(\frac{dy_1}{y_1} \wedge dy_2 / \frac{dx_1}{x_1} \wedge \frac{dx_2}{x_2}\right) - \nu_{G_1}(y_1) \\
&= i_o - a < i_o.
\end{aligned}$$

This completes the proof of the statement a).

b) We have only to consider the exceptional subcase $a \geq c+1$ in the case $i_o > i_1$. Take the standard neighborhood of q_1 with a system of regular parameters $(\frac{y_1}{y_2}, y_2)$. Then observing

$$\begin{cases} \nu_{G_1}(\frac{y_1}{y_2}) = a - c > 0 \\ \nu_{G_1}(y_2) = c > 0 \end{cases} \quad \& \quad \begin{cases} \nu_{E_p}(\frac{y_1}{y_2}) = a - (c+1) \geq 0 \\ \nu_{E_p}(y_2) = c + 1 > 0, \end{cases}$$

we conclude that the generic points of G_1 and E_p maps into the standard neighborhood of q_1.

Observe that the morphisms $f_0, \mathrm{bp}_1, \mathrm{cp}_{1,1}$ are all toroidal with respect to the modified logarithmic structures obtained by adding $\{y_2 = 0\}$ and $\{x_2 = 0\}$ (and their strict transforms) to the original boundary divisors. The original logarithmic structures coincide with the modified ones in neighborhoods of the generic points of G_1 and E_p and in the standard neighborhood of q_1. Therefore, $f_{1,1}$ is toroidal in neighborhoods of the generic points of G_1 and E_p with respect to the original logarithmic structures, and hence

$$\nu_{E_p}(R_{\log,f_1}) = 0$$

and

$$\nu_{G_1}(R_{\log,f_1}) = 0 < c = \nu_{G_1}(R_{\log,f_0}).$$

This proves statement a) and the first part of statement b).

The second part of statement b) follows inductively from the first part of the statement b) and Claim 3.3.3.B.2 b).

3.3.4) Subcase $2_p 1_q 0$: Neither G_1 nor G_2 maps onto q (and hence both G_1 and G_2 would map onto H_1)

This subcase does not happen. (Although this is again essentially due to Abhyankar's lemma, we will not use it explicitly). We will show by some elementary calculation that, in this subcase, $f|_{U_X}$ cannot be smooth and hence that we violate condition (c) imposed on a morphism in the logarithmic category (cf. 1.2.1).

First, we start with some ystems of regular parameters as described in 3.2. Then since $f^{-1}(D_Y) = D_X$ where $D_Y \cap U_q = H_1$ and $D_X \cap U_p = G_1 \cup G_2$, we have

$$f^* y_1 = u \cdot x_1^a x_2^b \quad \text{for some unit} \quad u \in \widehat{\mathcal{O}_{X,p}}^{\times} \quad \text{with} \quad a > 0, b > 0.$$

By replacing x_1 and x_2 with the ones multiplied by some appropriate units, we may assume

$$f^* y_1 = x_1^a x_2^b \quad \text{with} \quad a > 0, b > 0.$$

On the other hand, since neither G_1 nor G_2 maps onto q, neither x_1 nor x_2 divides $f^* y_2$ (Note that y_2 can be chosen in any way, irrelevant to the logarithmic structure, as long as (y_1, y_2) form a system of regular parameters.), in the Taylor expansion of $f^* y_2 = \Sigma \alpha_{ij} x_1^i x_2^j$ we conclude that there exists $i > 0$ such that $\alpha_{i0} \neq 0$ and that there exists $j > 0$ such that $\alpha_{0j} \neq 0$.

We set

$$i_o := \min\{i; \alpha_{i0} \neq 0\} > 0$$
$$j_o := \min\{i; \alpha_{0j} \neq 0\} > 0.$$

Finally we compute

$$\begin{aligned} f^*(dy_1 \wedge dy_2) &= df^* y_1 \wedge df^* y_2 \\ &= (a x_1^{a-1} x_2^b dx_1 + x_1^a b x_2^{b-1} dx_2) \\ &\wedge (\alpha_{i_o 0} i_o x_1^{i_o - 1} dx_1 + \Sigma_{i > i_o} \alpha_{i0} i x_1^{i-1} dx_1 \\ &\qquad + d(\Sigma_{i>0, j>0} \alpha_{ij} x_1^i x_2^j) \\ &\qquad + \Sigma_{j > j_o} \alpha_{0j} j x_2^{j-1} dx_2 + \alpha_{0 j_o} j_o x_2^{j_o - 1} dx_2) \\ &= (-b \alpha_{i_o 0} i_o x_1^{i_o} - \Sigma_{i > i_o} b \alpha_{i0} i x_1^{i - i_o} \cdot x_1^{i_o} \\ &\qquad + x_1 x_2 \cdot h \\ &\qquad + a \alpha_{0 j_o} j_o x_2^{j_o} + \Sigma_{j > j_o} a \alpha_{0j} j x_2^{j - j_o} \cdot x_2^{j_o}) \\ &\qquad \cdot x_1^{a-1} x_2^{b-1} dx_1 \wedge dx_2, \end{aligned}$$

noting that

$$(a x_1^{a-1} x_2^b dx_1 + x_1^a b x_2^{b-1} dx_2) \wedge d(\Sigma_{i>0, j>0} \alpha_{ij} x_1^i x_2^j)$$

is divisible by $x_1^a x_2^b \cdot dx_1 \wedge dx_2$ and can be written as

$$x_1 x_2 \cdot h \cdot x_1^{a-1} x_2^{b-1} dx_1 \wedge dx_2 \text{ for some } h \in \widehat{\mathcal{O}_{X,p}}.$$

But this implies that f ramifies along

$$\{(-b \alpha_{i_o 0} i_o x_1^{i_o} - \Sigma_{i > i_o} b \alpha_{i0} i x_1^{i - i_o} \cdot x_1^{i_o} + x_1 x_2 \cdot h + a \alpha_{0 j_o} j_o x_2^{j_o} + \Sigma_{j > j_o} a \alpha_{0j} j x_2^{j - j_o} \cdot x_2^{j_o}) = 0\}$$

other than possibly $\{x_1 = 0\}$ or $\{x_2 = 0\}$. This violates condition (c)

$$f|_{U_X = X - (\{x_1 = 0\} \cup \{x_2 = 0\})}$$

being smooth imposed on a morphism in the logarithmic category.

3.3.5) Subcase $2_p 1_q 1$: G_1 maps onto q but G_2 does not map onto q

3.3.5.A) Coordinate Expression

First, we start with some systems of regular parameters as described in 3.2. Then since $f^{-1}(D_Y) = D_X$ where $D_Y \cap U_q = H_1$ and $D_X \cap U_p = G_1 \cup G_2$, we have

$$f^* y_1 = u \cdot x_1^a x_2^b \quad \text{for some unit} \quad u \in \widehat{\mathcal{O}_{X,p}}^{\times} \quad \text{with} \quad a > 0, b > 0.$$

By replacing x_1 and x_2 with the ones multiplied by some appropriate units, we may assume

$$f^* y_1 = x_1^a x_2^b \quad \text{with} \quad a > 0, b > 0.$$

Set

$$f^* y_2 = \Sigma_{i>0,j>0,\det\begin{bmatrix} a & b \\ i & j \end{bmatrix}=0} \alpha_{ij} x_1^i x_2^j + \Sigma_{i>0,j\geq 0,\det\begin{bmatrix} a & b \\ i & j \end{bmatrix}\neq 0} \alpha_{ij} x_1^i x_2^j,$$

where no term of the form $x_2^j = x_1^0 x_2^j$ $(i = 0, j > 0)$ appears, since x_1 has to divide $f^* y_2$ as $G_1 = \{x_1 = 0\}$ maps onto q, and where there exists $i > 0$ with $\alpha_{i0} \neq 0$, since x_2 should not divide $f^* y_2$ as $G_2 = \{x_2 = 0\}$ does not map onto q.

Now since $f|_{U_X}$ must be smooth by condition (c) imposed on a morphism in the logarithmic category (cf. 1.2.1) and by computing the ramification $f^*(dy_1 \wedge dy_2)/dx_1 \wedge dx_2$ as in 3.3.2.A, we conclude that $f^* y_2$ must be of the form

$$\begin{aligned} f^* y_2 = & \Sigma_{(i,j)\geq(i_s,j_s),\det\begin{bmatrix} a & b \\ i & j \end{bmatrix}=0} \alpha_{ij} x_1^i x_2^j \\ & + \alpha_{i_o,0} x_1^{i_o} + \Sigma_{i\geq i_o>0,j\geq 0,(i,j)\neq(i_o,0),\det\begin{bmatrix} a & b \\ i & j \end{bmatrix}\neq 0} \alpha_{ij} x_1^i x_2^j, \end{aligned} \quad \text{with} \quad \alpha_{i_o,0} \neq 0,$$

where

$$i_0 = \min\{i; \alpha_{i0} \neq 0\}$$
$$(i_s, j_s) = \min\{(i,j); \alpha_{ij} \neq 0, \det \begin{bmatrix} a & b \\ i & j \end{bmatrix} = 0\}.$$

3.3.5.B) Descriptions of the canonical principalization cp_1 and the induced morphism f_1

Case $i_o \leq i_s$ and $i_o \leq a$: In this case, we have

$$\begin{cases} f^* y_1 = x_1^a x_2^b \\ f^* y_2 = v \cdot x_1^{i_o} \end{cases} \quad \text{with} \quad v \in \widehat{\mathcal{O}_{X,p}}^{\times} \quad \text{unit},$$

which implies $f^{-1}(y_1, y_2) \cdot \widehat{\mathcal{O}_{X,p}} = (x_1^{i_o})$ is principal. Therefore, cp_1 is an isomorphism in a neighborhood of p.

Diagram 3.3.5.B.1.

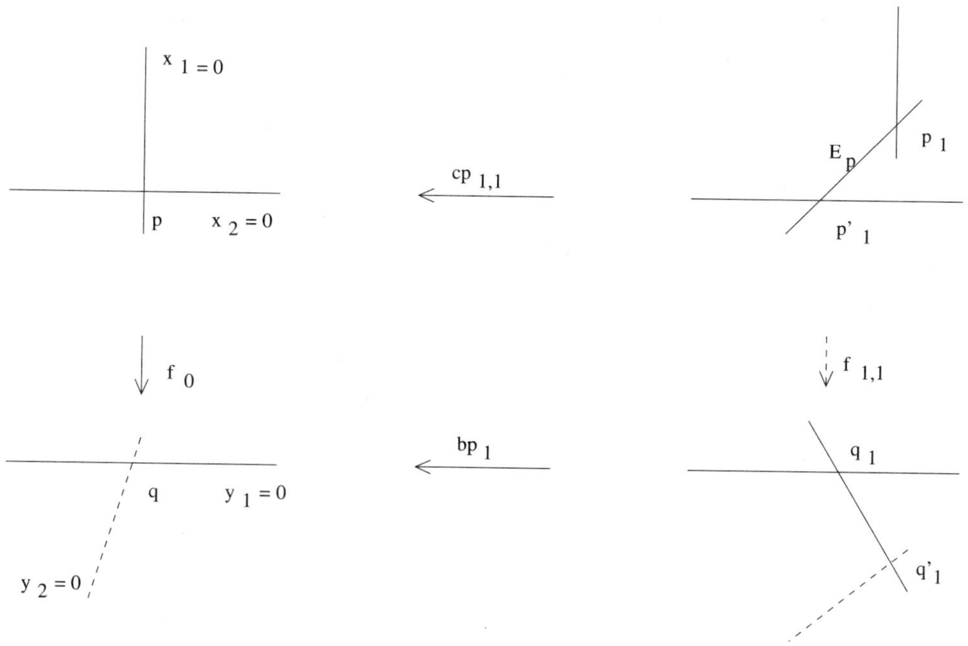

Case Otherwise, i.e., $i_o > i_s$ or $i_o > a$: We consider the 1st blowup $\mathrm{cp}_{1,1}$ of the canonical principalization cp_1, which is a sequence of blowups $\mathrm{cp}_1 = \mathrm{cp}_{1,1} \circ \cdots \circ \mathrm{cp}_{1,l}$ specified by the canonical principalization algorithm.

Claim 3.3.5.B.2. *The canonical principalization* cp_1 *is an isomorphism in a neighborhood of* p*, except under the case* $i_o > i_s$ *or* $i_o > a$*. In the exceptional case, we claim:*

a) The rational map $f_{1,1} = \mathrm{bp}^{-1} \circ f_0 \circ \mathrm{cp}_{1,1}$ *is well-defined (regular), that is to say, the ideal* $(f_0 \circ \mathrm{cp}_{1,1})^{-1}(y_1, y_2) \cdot \mathcal{O}_{X_{1,1}}$ *is principal, except possibly at finitely many points on* E_p.

b) At p_1*, the morphism* $f \circ \mathrm{cp}_{1,1}$ *is in Subcase* $2_{p_1}1_q2$*. At* p_1'*, the morphism* $f \circ \mathrm{cp}_{1,1}$ *is in Subcase* $2_{p_1'}1_q1$*. At any other point* $p_1''(\neq p_1, p_1')$ *on* E*, the morphism* $f \circ \mathrm{cp}_{1,1}$ *is in Subcase* $1_{p_1''}1_q1$*.*

Proof.

Statements a) and b) arc obvious.

3.3.5.C) Conclusion on the behavior of the logarithmic ramification divisor

Claim 3.3.5.C.1. *a) The coefficient of* G_1 *in the logarithmic ramification divisor strictly drops, while the coefficient of* G_2 *in the logarithmic ramification divisor remains 0. More precisely,*

$$\nu_{G_1}(R_{\log,f_1}) = \left\{ \begin{matrix} i_o - \min\{i_o, i_s\} & \text{if } a \geq \min\{i_o, i_s\} \\ i_o - a & \text{if } a < \min\{i_o, i_s\} \end{matrix} \right\} < i_o = \nu_{G_1}(R_{\log,f_0})$$

and

$$\nu_{G_2}(R_{\log,f_1}) = \nu_{G_2}(R_{\log,f_0}) = 0.$$

b) The canonical principalization is an isomorphism in a neighborhood of p, except under the case $i_o > i_s$ or $i_o > a$. In the exceptional case, the coefficient in $R_{\log,f_1}$ of the strict transform of the eceptional divisor E_p for $\mathrm{cp}_{1,1}$, obtained by blowing up p, is strictly smaller than the maximum of the coefficients of irreducible components in $R_{\log,f_0}$ in a neighborhood of p. More precisely,

$$\nu_{E_p}(R_{\log,f_1}) = \left.\begin{cases} i_o - \min\{i_o, i_s + j_s\} & \text{if } a+b \geq \min\{i_o, i_s+j_s\} \\ i_o - (a+b) & \text{if } a+b < \min\{i_o, i_s+j_s\} \end{cases}\right\} < i_o = \nu_{G_1}(R_{\log,f_0}).$$

More generally, the coefficient in $R_{\log,f_1}$, of any irreducible component E of the exceptional divisor for cp_1, is strictly smaller than the maximum of the coefficients of irreducible components in $R_{\log,f_0}$ in a neighborhood of p, i.e.,

$$\nu_E(R_{\log,f_1}) < i_o = \nu_{G_1}(R_{\log,f_0}).$$

Proof.

a) Firstly we compute the coefficient $\nu_{G_1}(R_{\log,f_0})$ of G_1 in the logarithmic ramification divisor $R_{\log,f_0}$:

$$\begin{aligned}
\nu_{G_1}(R_{\log,f_0}) &= \nu_{G_1}\left(\frac{dy_1}{y_1} \wedge dy_2 / \frac{dx_1}{x_1} \wedge \frac{dx_2}{x_2}\right) \\
&= \nu_{G_1}\left(\frac{d(x_1^a x_2^b)}{x_1^a x_2^b} \wedge d\left(\sum_{i \geq i_o > 0, j \geq 0, \det\begin{bmatrix} a & b \\ i & j \end{bmatrix} \neq 0} \alpha_{ij} x_1^i x_2^j\right) / \frac{dx_1}{x_1} \wedge \frac{dx_2}{x_2}\right) \\
&= \nu_{G_1}\left(\frac{d(x_1^a x_2^b)}{x_1^a x_2^b} \wedge d(x_1^{i_o}) / \frac{dx_1}{x_1} \wedge \frac{dx_2}{x_2}\right) = i_o.
\end{aligned}$$

Secondly we compute the coefficient of (the strict transform of) G_1 in $R_{\log,f_1}$.

Subcase $a \geq \min\{i_o, i_s\}$: In this subcase, we have

$$\begin{cases} \nu_{G_1}(\frac{y_1}{y_2}) = a - \min\{i_o, i_s\} \geq 0 \\ \nu_{G_1}(y_2) = \min\{i_o, i_s\} > 0, \end{cases}$$

which implies that the generic point of (the strict transform of) G_1 maps under f_1 into the standard neighborhood of q_1 with a system of regular parameters $(\frac{y_1}{y_2}, y_2)$. Therefore, we compute

$$\begin{aligned}
\nu_{G_1}(R_{\log,f_1}) &= \nu_{G_1}\left(\frac{d(\frac{y_1}{y_2})}{(\frac{y_1}{y_2})} \wedge \frac{dy_2}{y_2} / \frac{d(\frac{x_1}{x_2})}{(\frac{x_1}{x_2})} \wedge \frac{dx_2}{x_2}\right) \\
&= \nu_{G_1}\left(\frac{dy_1}{y_1} \wedge dy_2 / \frac{dx_1}{x_1} \wedge \frac{dx_2}{x_2}\right) - \nu_{G_1}(y_2) \\
&= i_o - \min\{i_o, i_s\} < i_o.
\end{aligned}$$

Subcase $a < \min\{i_o, i_s\}$: In this subcase, we have

$$\begin{cases} \nu_{G_1}(y_1) = a > 0 \\ \nu_{G_1}(\frac{y_2}{y_1}) = \min\{i_o, i_s\} - a > 0, \end{cases}$$

which implies that the generic point of (the strict transform of) G_1 maps under f_1 into the standard neighborhood of q_1' with a system of regular parameters $(y_1, \frac{y_2}{y_1})$. Therefore, we compute

$$\begin{aligned}\nu_{G_1}(R_{\log,f_1}) &= \nu_{G_1}\left(\frac{dy_1}{y_1}\wedge d(\frac{y_2}{y_1})/\frac{d(\frac{x_1}{x_2})}{(\frac{x_1}{x_2})}\wedge\frac{dx_2}{x_2}\right)\\ &= \nu_{G_1}\left(\frac{dy_1}{y_1}\wedge dy_2/\frac{dx_1}{x_1}\wedge\frac{dx_2}{x_2}\right) - \nu_{G_1}(y_1)\\ &= i_o - a < i_o.\end{aligned}$$

This completes the proof of statement a).

b) We have only to consider the exceptional case $i_o > i_s$ or $i_o > a$.

We compute the coefficient of (the strict transform of) the exceptional divisor E_p for $\mathrm{cp}_{1,1}$, obtained by blowing up p, in the logarithmic ramification divisor $R_{\log,f_1}$.

Subcase $a + b \geq \min\{i_o, i_s + j_s\}$: In this subcase, we have

$$\begin{cases}\nu_{E_p}(\frac{y_1}{y_2}) = a + b - \min\{i_o, i_s + j_s\} \geq 0\\ \nu_{E_p}(y_2) = \min\{i_o, i_s + j_s\} > 0,\end{cases}$$

which implies that the generic point of (the strict transform of) E_p maps under f_1 into the standard neighborhood of q_1 with a system of regular parameters $(\frac{y_1}{y_2}, y_2)$. Therefore, we compute

$$\begin{aligned}\nu_{E_p}(R_{\log,f_1}) &= \nu_{E_p}\left(\frac{d(\frac{y_1}{y_2})}{(\frac{y_1}{y_2})}\wedge\frac{dy_2}{y_2}/\frac{d(\frac{x_1}{x_2})}{(\frac{x_1}{x_2})}\wedge\frac{dx_2}{x_2}\right)\\ &= \nu_{E_p}\left(\frac{dy_1}{y_1}\wedge dy_2/\frac{dx_1}{x_1}\wedge\frac{dx_2}{x_2}\right) - \nu_{E_p}(y_2)\\ &= i_o - \min\{i_o, i_s + j_s\} < i_o.\end{aligned}$$

Subcase $a + b < \min\{i_o, i_s + j_s\}$: In this subcase, we have

$$\begin{cases}\nu_{E_p}(y_1) = a + b > 0\\ \nu_{E_p}(\frac{y_2}{y_1}) = \min\{i_o, i_s + j_s\} - (a + b) > 0,\end{cases}$$

which implies that the generic point of (the strict transform of) E_p maps under f_1 into the standard neighborhood of q_1' with a system of regular parameters $(y_1, \frac{y_2}{y_1})$. Therefore, we compute

$$\begin{aligned}\nu_{E_p}(R_{\log,f_1}) &= \nu_{E_p}\left(\frac{dy_1}{y_1}\wedge d(\frac{y_2}{y_1})/\frac{d(\frac{x_1}{x_2})}{(\frac{x_1}{x_2})}\wedge\frac{dx_2}{x_2}\right)\\ &= \nu_{E_p}\left(\frac{dy_1}{y_1}\wedge dy_2/\frac{dx_1}{x_1}\wedge\frac{dx_2}{x_2}\right) - \nu_{E_p}(y_1)\\ &= i_o - (a + b) < i_o.\end{aligned}$$

This completes the proof of the first part of statement b).

In order to verify the second part of statement b), we look at the points where $(f \circ \mathrm{cp}_{1,1})^{-1}(y_1, y_2) \cdot \mathcal{O}_{X_{1,1}}$ is (possibly) not principal.

At p_1, the morphism $f \circ \mathrm{cp}_{1,1}$ is in Subcase $2_{p_1} 1_q 2$ with

$$\nu_{G_1}(R_{\log, f \circ \mathrm{cp}_{1,1}}) - i_o$$
$$\nu_{E_p}(R_{\log, f \circ \mathrm{cp}_{1,1}}) = i_o.$$

Therefore, by Claim 3.3.6.C.1 b) we have for any irreducible component E, other than (the strict transform of) E_p, which is exceptional for cp_1 and which maps onto p_1 under $\mathrm{cp}_{1,2} \circ \cdots \circ \mathrm{cp}_{1,l}$

$$\nu_E(R_{\log, f_1}) < \max\{\nu_{G_1}(R_{\log, f \circ \mathrm{cp}_{1,1}}), \nu_{E_p}(R_{\log, f \circ \mathrm{cp}_{1,1}})\} = i_o = \nu_{G_1}(R_{\log, f_0}).$$

At p_1', the morphism $f \circ \mathrm{cp}_{1,1}$ is in Subcase $2_{p_1'} 1_q 1$ with

$$\nu_{E_p}(R_{\log, f \circ \mathrm{cp}_{1,1}}) = i_o.$$

Therefore, by the first part of statement b) and by induction on the length of the sequence of blowups of points for cp_1, we have for any irreducible component E, other than (the strict transform of) E_p, which is exceptional for cp_1 and which maps onto p_1' under $\mathrm{cp}_{1,2} \circ \cdots \circ \mathrm{cp}_{1,l}$

$$\nu_E(R_{\log, f_1}) < \nu_{E_p}(R_{\log, f \circ \mathrm{cp}_{1,1}}) = i_o = \nu_{G_1}(R_{\log, f_0}).$$

At $p_1'' \in E_p$, other than p_1 or p_1', the morphism $(f \circ \mathrm{cp}_{1,1})$ is in Subcase $1_{p_1''} 1_q 1$. Therefore, by Claim 3.3.3.C.1 b) we have for any irreducible component E, other than (the strict transform of) E_p, which is exceptional for cp_1 and which maps onto p_1'' under $\mathrm{cp}_{1,2} \circ \cdots \circ \mathrm{cp}_{1,l}$

$$\nu_E(R_{\log, f_1}) = 0 < \nu_{E_p}(R_{\log, f \circ \mathrm{cp}_{1,1}}) = i_o = \nu_{G_1}(R_{\log, f_0}).$$

3.3.6) Subcase $2_p 1_q 2$: Both G_1 and G_2 map onto q

3.3.6.A) Coordinate Expression

First, we start with some systems of regular parameters as described in 3.2. Then since $f^{-1}(D_Y) = D_X$ where $D_Y \cap U_q = H_1$ and $D_X \cap U_p = G_1 \cup G_2$, we have

$$f^* y_1 = u \cdot x_1^a x_2^b \quad \text{for some unit} \quad u \in \widehat{\mathcal{O}_{X,p}}^{\times} \quad \text{with} \quad a > 0, b > 0.$$

By replacing x_1 and x_2 with the ones multiplied by some appropriate units, we may assume

$$f^* y_1 = x_1^a x_2^b \quad \text{with} \quad a > 0, b > 0.$$

Set

$$f^* y_2 = \Sigma_{i>0, j>0, \det \begin{bmatrix} a & b \\ i & j \end{bmatrix} = 0} \alpha_{ij} x_1^i x_2^j + \Sigma_{i>0, j>0, \det \begin{bmatrix} a & b \\ i & j \end{bmatrix} \neq 0} \alpha_{ij} x_1^i x_2^j,$$

where $x_1 x_2$ divides $f^* y_2$, since both G_1 and G_2 maps onto q.

Now since $f|_{U_X}$ must be smooth by condition (c) imposed on a morphism in the logarithmic category (cf. 1.2.1) and by computing the ramification $f^*(dy_1 \wedge dy_2)/dx_1 \wedge dx_2$ as in 3.3.2.A, we conclude that $f^* y_2$ must be of the form

$$\begin{aligned} f^* y_2 = &\Sigma_{(i,j) \geq (i_s, j_s), \det \begin{bmatrix} a & b \\ i & j \end{bmatrix} = 0} \alpha_{ij} x_1^i x_2^j \\ &+ \alpha_{i_o, j_o} x_1^{i_o} x_2^{j_o} + \Sigma_{i \geq i_o > 0, j \geq j_o > 0, (i,j) \neq (i_o, j_o), \det \begin{bmatrix} a & b \\ i & j \end{bmatrix} \neq 0} \alpha_{ij} x_1^i x_2^j, \end{aligned} \quad \text{with } \alpha_{i_o, j_o} \neq 0,$$

where

$$\begin{aligned} i_o &= \min\{i; \alpha_{ij} \neq 0, \det \begin{bmatrix} a & b \\ i & j \end{bmatrix} \neq 0\} \\ j_o &= \min\{j; \alpha_{ij} \neq 0, \det \begin{bmatrix} a & b \\ i & j \end{bmatrix} \neq 0\} \\ &\text{with } \det \begin{bmatrix} a & b \\ i_o & j_o \end{bmatrix} \neq 0, \text{ and} \\ (i_s, j_s) &= \min\{(i,j); \alpha_{ij} \neq 0, \det \begin{bmatrix} a & b \\ i & j \end{bmatrix} = 0\}. \end{aligned}$$

3.3.6.B) Descriptions of the canonical principalization cp_1 and the induced morphism f_1

Case $\left\{ \begin{matrix} i_o \geq i_s \\ j_o \geq j_s \end{matrix} \right\}$ or $\left\{ \begin{matrix} i_o \leq i_s & & i_o \leq a \\ & \& & \\ j_o \leq j_s & & j_o \leq b \end{matrix} \right\}$ or $\left\{ \begin{matrix} i_o \leq i_s & & i_o \geq a \\ & \& & \\ j_o \leq j_s & & j_o \geq b \end{matrix} \right\}$:

In this case, the ideal

$$\begin{aligned} f^{-1}(y_1, y_2) \cdot \widehat{\mathcal{O}_{X,p}} = (x_1^{i_s} x_2^{j_s}) \ &\text{or} \ (x_1^a x_2^b), \\ &\text{or} \\ &(x_1^{i_o} x_2^{j_o}), \\ &\text{or} \\ &(x_1^a x_2^b), \ \text{respectively}, \end{aligned}$$

is principal. Therefore, cp_1 is an isomorphism in a neighborhood of p.

Case Otherwise:

We consider the 1st blowup $\mathrm{cp}_{1,1}$ of the canonical principalization cp_1, which is a sequence of blowups $\mathrm{cp}_1 = \mathrm{cp}_{1,1} \circ \cdots \circ \mathrm{cp}_{1,l}$ specified by the canonical principalization algorithm.

Diagram 3.3.6.B.1.

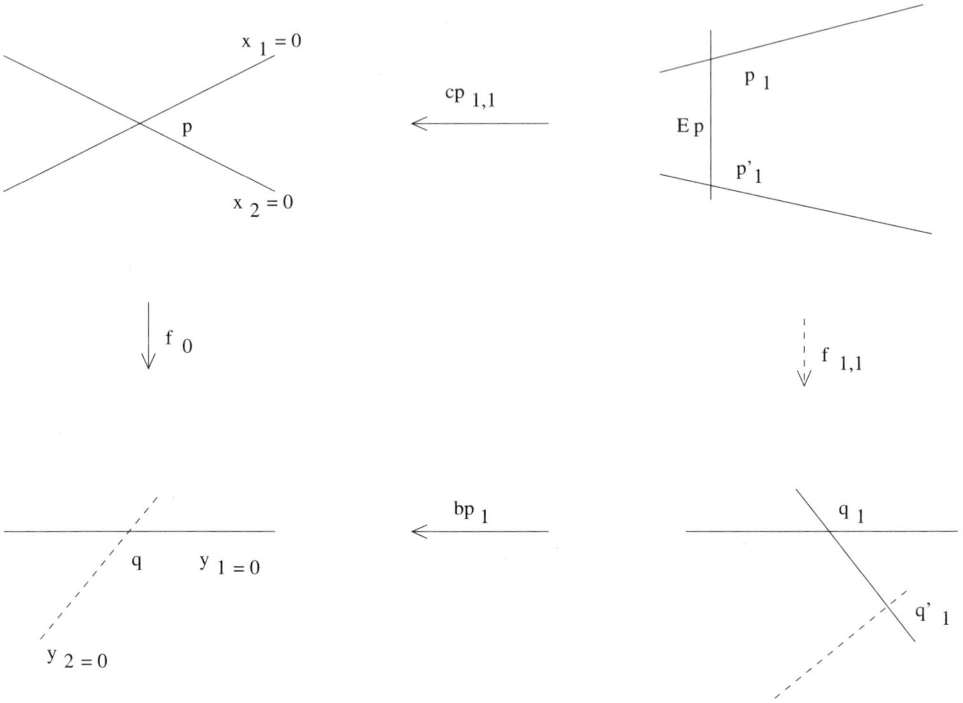

Claim 3.3.6.B.2. *The canonical principalization is an isomorphism in a neighborhood of p, except for the case where $f^{-1}(y_1, y_2) \cdot \widehat{\mathcal{O}_{X,p}}$ is not principal. In the exceptional case, we claim:*

a) The rational map $f_{1,1} = \mathrm{bp}^{-1} \circ f_0 \circ \mathrm{cp}_{1,1}$ is well-defined (regular), that is to say, the ideal $(f_0 \circ \mathrm{cp}_{1,1})^{-1}(y_1, y_2) \cdot \mathcal{O}_{X_{1,1}}$ is principal, except possibly at finitely many points on E_p.

b) At p_1 (resp. p_1'), the morphism $f \circ \mathrm{cp}_{1,1}$ is in Subcase $2_{p_1}1_q1$ (resp. $2_{p_1'}1_q1$). At any other point p_1'' on E_p, the morphism $f \circ \mathrm{cp}_{1,1}$ is in Subcase $1_{p_1''}1_q1$.

Proof.

Statements a) and b) are obvious.

3.3.6.C) Conclusion on the behavior of the logarithmic ramification divisor

Claim 3.3.6.C.1. *a) The coefficients of G_1 and G_2 in the logarithmic ramification*

divisor strictly drop. More precisely,

$$\nu_{G_1}(R_{\log,f_1}) = \left\{ \begin{matrix} i_o - \min\{i_o, i_s\} & \text{if } a \geq \min\{i_o, i_s\} \\ i_o - a & \text{if } a < \min\{i_o, i_s\} \end{matrix} \right\} < i_o = \nu_{G_1}(R_{\log,f_0})$$

$$\nu_{G_2}(R_{\log,f_1}) = \left\{ \begin{matrix} j_o - \min\{j_o, j_s\} & \text{if } b \geq \min\{j_o, j_s\} \\ j_o - b & \text{if } b < \min\{j_o, j_s\} \end{matrix} \right\} < j_o = \nu_{G_2}(R_{\log,f_0}).$$

b) The canonical principalization is an isomorphism in a neighborhood of p*, except for the case when* $f^{-1}(y_1, y_2) \cdot \widehat{\mathcal{O}_{X,p}}$ *is not principal. In the exceptional case where* $f^{-1}(y_1, y_2) \cdot \widehat{\mathcal{O}_{X,p}}$ *is not principal, the coefficient in* $R_{\log,f_1}$ *of the strict transform of the eceptional divisor* E_p *for* $\mathrm{cp}_{1,1}$*, obtained by blowing up* p*, is strictly smaller than the maximum of the coefficients of the irreducible components in* $R_{\log,f_0}$ *in a neighborhood of* p*. More precisely,*

$$\nu_{E_p}(R_{\log,f_1}) = \left\{ \begin{matrix} i_o + j_o - \min\{i_o + j_o, i_1 + j_1\} & \text{if } a + b \geq \min\{i_o + j_o, i_1 + j_1\} \\ i_o + j_o - (a + b) & \text{if } a + b < \min\{i_o + j_o, i_1 + j_1\} \end{matrix} \right\}$$
$$< \max\{i_o, j_o\} = \max\{\nu_{G_1}(R_{\log,f_0}), \nu_{G_2}(R_{\log,f_0})\}.$$

More generally, the coefficient in $R_{\log,f_1}$*, of any irreducible component* E *of the exceptional divisor for* cp_1*, is strictly smaller than the maximum of the coefficients of the irreducible components in* $R_{\log,f_0}$ *in a neighborhood of* p*, i.e.,*

$$\nu_E(R_{\log,f_1}) < \max\{\nu_{G_1}(R_{\log,f_0}), \nu_{G_1}(R_{\log,f_0})\} = \max\{i_o, j_o\}.$$

Proof.

a) Firstly we compute the coefficient $\nu_{G_1}(R_{\log,f_0})$ of G_1 in the logarithmic ramification divisor $R_{\log,f_0}$.

$$\begin{aligned} \nu_{G_1}(R_{\log,f_0}) &= \nu_{G_1}\left(\frac{dy_1}{y_1} \wedge dy_2 / \frac{dx_1}{x_1} \wedge \frac{dx_2}{x_2}\right) \\ &= \nu_{G_1}\left(\frac{d(x_1^a x_2^b)}{x_1^a x_2^b} \wedge d\left(\Sigma_{i \geq i_o > 0, j \geq j_o > 0, \det\begin{bmatrix} a & b \\ i & j \end{bmatrix} \neq 0} \alpha_{ij} x_1^i x_2^j\right) / \frac{dx_1}{x_1} \wedge \frac{dx_2}{x_2}\right) \\ &= \nu_{G_1}\left(\frac{d(x_1^a x_2^b)}{x_1^a x_2^b} \wedge d(x_1^{i_o} x_2^{j_o}) / \frac{dx_1}{x_1} \wedge \frac{dx_2}{x_2}\right) = i_o. \end{aligned}$$

Symmetrically, we compute the coefficient $\nu_{G_2}(R_{\log,f_0})$ of G_2 in the logarithmic ramification divisor $R_{\log,f_0}$ to be

$$\nu_{G_2}(R_{\log,f_0}) = j_o.$$

Secondly we compute the coefficient of (the strict transform of) G_1 in $R_{\log,f_1}$.

Subcase $a \geq \min\{i_o, i_s\}$: In this subcase, we have

$$\left\{ \begin{matrix} \nu_{G_1}(\frac{y_1}{y_2}) = a - \min\{i_o, i_s\} \geq 0 \\ \nu_{G_1}(y_2) = \min\{i_o, i_s\} > 0, \end{matrix} \right.$$

which implies that the generic point of (the strict transform of) G_1 maps under f_1 into the standard neighborhood of q_1 with a system of regular parameters $(\frac{y_1}{y_2}, y_2)$. Therefore, we compute

$$\begin{aligned}\nu_{G_1}(R_{\log,f_1}) &= \nu_{G_1}\left(\frac{d(\frac{y_1}{y_2})}{(\frac{y_1}{y_2})} \wedge \frac{dy_2}{y_2} / \frac{d(\frac{x_1}{x_2})}{(\frac{x_1}{x_2})} \wedge \frac{dx_2}{x_2}\right) \\ &= \nu_{G_1}\left(\frac{dy_1}{y_1} \wedge dy_2 / \frac{dx_1}{x_1} \wedge \frac{dx_2}{x_2}\right) - \nu_{G_1}(y_2) \\ &= i_o - \min\{i_o, i_s\} < i_o.\end{aligned}$$

Subcase $a < \min\{i_o, i_s\}$: In this subcase, we have

$$\begin{cases} \nu_{G_1}(y_1) = a > 0 \\ \nu_{G_1}(\frac{y_2}{y_1}) = \min\{i_o, i_s\} - a > 0, \end{cases}$$

which implies that the generic point of (the strict transform of) G_1 maps under f_1 into the standard neighborhood of q_1' with a system of regular parameters $(y_1, \frac{y_2}{y_1})$. Therefore, we compute

$$\begin{aligned}\nu_{G_1}(R_{\log,f_1}) &= \nu_{G_1}\left(\frac{dy_1}{y_1} \wedge d(\frac{y_2}{y_1}) / \frac{d(\frac{x_1}{x_2})}{(\frac{x_1}{x_2})} \wedge \frac{dx_2}{x_2}\right) \\ &= \nu_{G_1}\left(\frac{dy_1}{y_1} \wedge dy_2 / \frac{dx_1}{x_1} \wedge \frac{dx_2}{x_2}\right) - \nu_{G_1}(y_1) \\ &= i_o - a < i_o.\end{aligned}$$

Symmetrically, we compute

$$\nu_{G_2}(R_{\log,f_1}) = \left\{\begin{matrix} j_o - \min\{j_o, j_s\} & \text{if } \ b \geq \min\{j_o, j_s\} \\ j_o - b & \text{if } \ b < \min\{j_o, j_s\} \end{matrix}\right\} < j_o = \nu_{G_2}(R_{\log,f_0}).$$

This proves statement a).

b) We compute the coefficient of (the strict transform of) the exceptional divisor E_p for $\mathrm{cp}_{1,1}$, obtained by blowing up p, in the logarithmic ramification divisor $R_{\log,f_1}$.

Subcase $a + b \geq \min\{i_o + j_o, i_s + j_s\}$: In this subcase, we have

$$\begin{cases} \nu_{E_p}(\frac{y_1}{y_2}) = a + b - \min\{i_o + j_o, i_s + j_s\} \geq 0 \\ \nu_{E_p}(y_2) = \min\{i_o + j_o, i_s + j_s\} > 0, \end{cases}$$

which implies that the generic point of (the strict transform of) E_p maps under f_1 into the standard neighborhood of q_1 with a system of regular parameters $(\frac{y_1}{y_2}, y_2)$. Therefore, we compute

$$\begin{aligned}\nu_{E_p}(R_{\log,f_1}) &= \nu_{E_p}\left(\frac{d(\frac{y_1}{y_2})}{(\frac{y_1}{y_2})} \wedge \frac{dy_2}{y_2} / \frac{d(\frac{x_1}{x_2})}{(\frac{x_1}{x_2})} \wedge \frac{dx_2}{x_2}\right) \\ &= \nu_{E_p}\left(\frac{dy_1}{y_1} \wedge dy_2 / \frac{dx_1}{x_1} \wedge \frac{dx_2}{x_2}\right) - \nu_{E_p}(y_2) \\ &= i_o + j_o - \min\{i_o + j_o, i_s + j_s\}.\end{aligned}$$

Now we claim that

$$i_o + j_o - \min\{i_o + j_o, i_s + j_s\} < \max\{i_o, j_o\}$$

unless $f^{-1}(y_1, y_2) \cdot \widehat{\mathcal{O}_{X,p}}$ is already principal.

In fact, if

$$i_o + j_o \leq i_s + j_s,$$

then

$$i_o + j_o - \min\{i_o + j_o, i_s + j_s\} = 0 < \max\{i_o, j_o\}.$$

Therefore, we may assume

$$i_o + j_o > i_s + j_s,$$

in which case we have

$$\begin{aligned} i_o + j_o - \min\{i_o + j_o, i_s + j_s\} &= (i_o + j_o) - (i_s + j_s) \\ &= (i_o - i_s) + (j_o - j_s) \\ &= \begin{cases} (i_o - i_s) - (j_s - j_o) < i_o & \text{if } j_s \geq j_o \\ (j_o - j_s) - (i_s - i_o) < j_o & \text{if } i_s \geq i_o \\ \text{or} \\ f^{-1}(y_1, y_2) \cdot \widehat{\mathcal{O}_{X,p}} = (x_1^a x_2^b, x_1^{i_s} x_2^{j_s}) = (x_1^{\min\{a, i_s\}} x_2^{\min\{b, j_s\}}) \\ \qquad\qquad \text{is principal if } i_s < i_o \ \&\ j_s < j_o. \end{cases} \end{aligned}$$

Subcase $a + b < \min\{i_o + j_o, i_s + j_s\}$: In this subcase, we have

$$\begin{cases} \nu_{E_p}(y_1) = a + b > 0 \\ \nu_{E_p}(\frac{y_2}{y_1}) = \min\{i_o + j_o, i_s + j_s\} - (a + b) > 0, \end{cases}$$

which implies that the generic point of (the strict transform of) E_p maps under f_1 into the standard neighborhood of q_1' with a system of regular parameters $(y_1, \frac{y_2}{y_1})$. Therefore, we compute

$$\begin{aligned} \nu_{E_p}(R_{\log, f_1}) &= \nu_{E_p}\left(\frac{dy_1}{y_1} \wedge d(\frac{y_2}{y_1}) / \frac{d(\frac{x_1}{x_2})}{(\frac{x_1}{x_2})} \wedge \frac{dx_2}{x_2}\right) \\ &= \nu_{E_p}\left(\frac{dy_1}{y_1} \wedge dy_2 / \frac{dx_1}{x_1} \wedge \frac{dx_2}{x_2}\right) - \nu_{E_p}(y_1) \\ &= i_o + j_o - (a + b). \end{aligned}$$

Now we claim that

$$i_o + j_o - (a + b) < \max\{i_o, j_o\}$$

unless $f^{-1}(y_1, y_2) \cdot \widehat{\mathcal{O}_{X,p}}$ is already principal.

In fact, we have

$$i_o + j_o - (a+b) = (i_o - a) + (j_o - b)$$

$$= \begin{cases} (i_o - a) - (b - j_o) < i_o & \text{if } b \geq j_o \\ (j_o - b) - (a - i_o) < j_o & \text{if } a \geq i_o \\ \text{or} \\ f^{-1}(y_1, y_2) \cdot \widehat{\mathcal{O}_{X,p}} = (x_1^a x_2^b) \\ \qquad\qquad \text{principal} \ \text{if } a < i_o \ \& \ b < j_o \\ (\text{Note that } a < i_s \ \& \ b < j_s \\ \ \text{by the subcase assumption } a + b < \min\{i_o + j_o, i_s + j_s\}.) \end{cases}$$

This proves the first part of statement b).

It remains to prove the second part of statement b).

Suppose $p_k \in D_{X_{1,k}} \subset X_{1,k}$ is a closed point sitting over $p = p_0 \in D_X = D_{X_{1,0}} \subset X = X_{1,0}$, i.e.,

$$\mathrm{cp}_{1,1} \circ \cdots \circ \mathrm{cp}_{1,k}(p_k) = p.$$

We set

$$g_k = f \circ (\mathrm{cp}_{1,1} \circ \cdots \circ \mathrm{cp}_{1,k}).$$

Suppose that $g_k^{-1}(y_1, y_2) \cdot \widehat{\mathcal{O}_{X_{1,k},p_k}}$ is not principal, and hence that we blowup p_k to obtain the exceptional divisor E_{p_k} for $\mathrm{cp}_{1,k+1}$, the $(k+1)$-th stage of the canonical principalization $\mathrm{cp}_1 = \mathrm{cp}_{1,1} \circ \cdots \circ \mathrm{cp}_{1,k} \circ \mathrm{cp}_{1,k+1} \circ \cdots \circ \mathrm{cp}_{1,l}$.

At p_k, the morphism g_k is either in Subcase $1_{p_k}1_q1$ or in Subcase $2_{p_k}1_q2$.

If g_k is in Subcase $1_{p_k}1_q1$ at p_k, then by Claim 3.3.3.C.1 we have

$$\nu_E(R_{\log, f_1}) = 0 < \max\{i_o, j_o\}$$

for (the strict transform of) any exceptional divisor E for $\mathrm{cp}_{1,k+1} \circ \cdots \circ \mathrm{cp}_{1,l}$ which maps onto p_k.

In order to analyze the case where g_k is in Subcase $2_{p_k}1_q2$, we prove the following inductive lemma.

Lemma 3.3.6.C.2. *Let $p_k \in D_{X_{1,k}} \subset X_{1,k}$ be a point, sitting over p, where g_k is in Subcase $2_{p_k}1_q2$. Assume that the ideal $g_k^{-1}(y_1, y_2) \cdot \widehat{\mathcal{O}_{X_{1,k},p_k}}$ is not principal.*

Suppose that we have the following two conditions:

$(i)_k$ There exists a system of regular parameters $(x_{1,k}, x_{2,k})$ of $\widehat{\mathcal{O}_{X_{1,k},p_k}}$, compatible with the logarithmic structure, such that

$$(*)_k \begin{cases} g_k^* y_1 = x_{1,k}^{a_k} x_{2,k}^{b_k} \quad \text{with} \quad a_k > 0, b_k > 0 \\ g_k^* y_2 = \Sigma_{(i,j) \geq (i_{s,k}, j_{s,k}), \det \begin{bmatrix} a_k & b_k \\ i & j \end{bmatrix} = 0} \alpha_{ij} x_{1,k}^i x_{2,k}^j \\ \qquad + \alpha_{i_{o,k}, j_{o,k}} x_{1,k}^{i_{o,k}} x_{2,k}^{j_{o,k}} + \Sigma_{i \geq i_{o,k} > 0, j \geq j_{o,k} > 0, (i,j) \neq (i_{o,k}, j_{o,k}), \det \begin{bmatrix} a_k & b_k \\ i & j \end{bmatrix} \neq 0} \alpha_{ij} x_{1,k}^i x_{2,k}^j \\ \qquad \text{with} \quad \alpha_{i_{o,k}, j_{o,k}} \neq 0, \end{cases}$$

where

$$i_{o,k} = \min\{i; \alpha_{ij} \neq 0, \det \begin{bmatrix} a_k & b_k \\ i & j \end{bmatrix} \neq 0\}$$

$$j_{o,k} = \min\{j; \alpha_{ij} \neq 0, \det \begin{bmatrix} a_k & b_k \\ i & j \end{bmatrix} \neq 0\}$$

$$\textit{with } \det \begin{bmatrix} a_k & b_k \\ i_{o,k} & j_{o,k} \end{bmatrix} \neq 0, \textit{ and}$$

$$(i_{s,k}, j_{s,k}) = \min\{(i,j); \alpha_{ij} \neq 0, \det \begin{bmatrix} a_k & b_k \\ i & j \end{bmatrix} = 0\}.$$

$(ii)_k$ *One of the following four holds:*

$$(\alpha)_k \begin{Bmatrix} i_{o,k} > i_{s,k} \\ j_{o,k} < j_{s,k} \end{Bmatrix} \quad \textit{with } \ i_{o,k} - i_{s,k} < i_o \ \textit{ and } \ i_{o,k} - a_k < i_o$$

$$(\beta)_k \begin{Bmatrix} i_{o,k} < i_{s,k} \\ j_{o,k} > j_{s,k} \end{Bmatrix} \quad \textit{with } \ j_{o,k} - j_{s,k} < j_o \ \textit{ and } \ j_{o,k} - b_k < j_o$$

$$(\gamma)_k \begin{Bmatrix} i_{o,k} \leq i_{s,k} & & i_{o,k} > a_k \\ & \& & \\ j_{o,k} \leq j_{s,k} & & j_{o,k} < b_k \end{Bmatrix} \quad \textit{with } \ i_{o,k} - a_k < i_o,$$

$$(\delta)_k \begin{Bmatrix} i_{o,k} \leq i_{s,k} & & i_{o,k} < a_k \\ & \& & \\ j_{o,k} \leq j_{s,k} & & j_{o,k} > b_k \end{Bmatrix} \quad \textit{with } \ j_{o,k} - b_k < j_o.$$

Then, E_{p_k} being the exceptional divisor for $\mathrm{cp}_{1,k+1}$, obtained by blowing up p_k, we have

$$\nu_{E_{p_k}}(R_{\log,f_1}) < \max\{i_o, j_o\}.$$

Moreover, let $p_{k+1} \in D_{X_{1,k+1}} \subset X_{1,k+1}$ be a point over p_k where g_{k+1} is in Subcase $2_{p_{k+1}}1_q2$. Assume $g_{k+1}^{-1}(y_1, y_2) \cdot \widehat{\mathcal{O}_{X_{1,k+1},p_{k+1}}}$ is not principal. Then we have the conditions $(i)_{k+1}$ with coordinate expression $(*)_{k+1}$ and $(ii)_{k+1}$ (derived inductively from the conditions $(i)_k$ with coordinate expression $(*)_k$ and $(ii)_k$).

(Note that in the special case where in $g_0^* y_2$ there is no term $x_{1,0}^i x_{2,0}^j$ with $\det \begin{bmatrix} a_0 & b_0 \\ i & j \end{bmatrix} = 0$ and hence where subsequently in $g_k^* y_2$ there is no term $x_{1,k}^i x_{2,k}^j$ with $\det \begin{bmatrix} a_k & b_k \\ i & j \end{bmatrix} = 0$, we set $i_{s,0} = i_{s,k} = j_{s,0} = j_{s,k} = \infty$ by convention.)

Proof.

We compute the coefficient $\nu_{E_{p_k}}(R_{\log,f_1})$ of (the strict transform of) the exceptional divisor E_{p_k} for $\mathrm{cp}_{1,k+1}$, obtained by blowing up p_k, in the logarithmic ramification divisor $R_{\log,f_1}$. (We note that the computation is identical to the one given for $\nu_{E_p}(R_{\log,f_1}) = \nu_{E_{p_0}}(R_{\log,f_1})$, except that at the end we compare $\nu_{E_{p_k}}(R_{\log,f_1})$ to $\max\{i_o, j_o\}$ but not to $\max\{i_{o,k}, j_{o,k}\}$.).

Diagram 3.3.6.C.3.

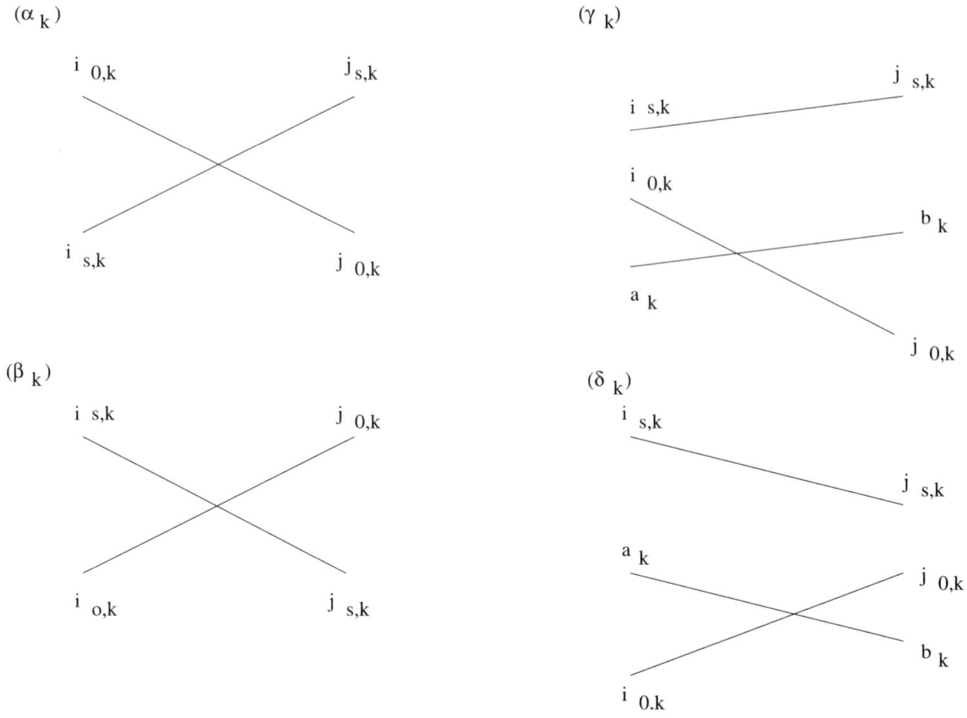

Subcase $a_k + b_k \geq \min\{i_{o,k} + j_{o,k}, i_{s,k} + j_{s,k}\}$: In this subcase, we have

$$\begin{cases} \nu_{E_{p_k}}(\frac{y_1}{y_2}) = a_k + b_k - \min\{i_{o,k} + j_{o,k}, i_{s,k} + j_{s,k}\} \geq 0 \\ \nu_{E_{p_k}}(y_2) = \min\{i_{o,k} + j_{o,k}, i_{s,k} + j_{s,k}\} > 0, \end{cases}$$

which implies that the generic point of (the strict transform of) E_{p_k} maps under f_1 into the standard neighborhood of q_1 with a system of regular parameters $(\frac{y_1}{y_2}, y_2)$. Therefore, we compute

$$\begin{aligned} \nu_{E_{p_k}}(R_{\log, f_1}) &= \nu_{E_{p_k}} \left(\frac{d(\frac{y_1}{y_2})}{(\frac{y_1}{y_2})} \wedge \frac{dy_2}{y_2} / \frac{d(\frac{x_{1,k}}{x_{2,k}})}{(\frac{x_{1,k}}{x_{2,k}})} \wedge \frac{dx_{2,k}}{x_{2,k}} \right) \\ &= \nu_{E_{p_k}} \left(\frac{dy_1}{y_1} \wedge dy_2 / \frac{dx_{1,k}}{x_{1,k}} \wedge \frac{dx_{2,k}}{x_{2,k}} \right) - \nu_{E_{p_k}}(y_2) \\ &= i_{o,k} + j_{o,k} - \min\{i_{o,k} + j_{o,k}, i_{s,k} + j_{s,k}\}. \end{aligned}$$

Now we claim that

$$i_{o,k} + j_{o,k} - \min\{i_{o,k} + j_{o,k}, i_{s,k} + j_{s,k}\} < \max\{i_o, j_o\}$$

unless $g_k^{-1}(y_1, y_2) \cdot \widehat{\mathcal{O}_{X_{1,k}, p_k}}$ is already principal.

In fact, if

$$i_{o,k} + j_{o,k} \leq i_{s,k} + j_{s,k},$$

then

$$i_{o,k} + j_{o,k} - \min\{i_{o,k} + j_{o,k}, i_{s,k} + j_{s,k}\} = 0 < \max\{i_o, j_o\}.$$

Therefore, we may assume

$$i_{o,k} + j_{o,k} > i_{s,k} + j_{s,k},$$

in which case we have

$$\begin{aligned} i_{o,k} + j_{o,k} - \min\{i_{o,k} + j_{o,k}, i_{s,k} + j_{s,k}\} &= (i_{o,k} + j_{o,k}) - (i_{s,k} + j_{s,k}) \\ &= (i_{o,k} - i_{s,k}) + (j_{o,k} - j_{s,k}) \\ &= \begin{cases} (i_{o,k} - i_{s,k}) - (j_{s,k} - j_{o,k}) < i_o \text{ if } j_{s,k} \geq j_{o,k} \\ (j_{o,k} - j_{s,k}) - (i_{s,k} - i_{o,k}) < j_o \text{ if } i_{s,k} \geq i_{o,k} \\ \text{or} \\ g_k^{-1}(y_1, y_2) \cdot \widehat{\mathcal{O}_{X_{1,k},p_k}} = (x_{1,k}^{a_k} x_{2,k}^{b_k}, x_{1,k}^{i_{s,k}} x_{2,k}^{j_{s,k}}) = (x_{1,k}^{\min\{a_k,i_{s,k}\}} x_{2,k}^{\min\{b_k,j_{s,k}\}}) \\ \qquad\qquad \text{principal if } i_{s,k} < i_{o,k} \ \& \ j_{s,k} < j_{o,k}. \end{cases} \end{aligned}$$

Remark that, under the assumption

$$i_{o,k} + j_{o,k} > i_{s,k} + j_{s,k},$$

when $g_k^{-1}(y_1, y_2) \cdot \widehat{\mathcal{O}_{X_{1,k},p_k}}$ is not principal, we have

$$\begin{aligned} j_{s,k} \geq j_{o,k} &\Longrightarrow (\alpha)_k \quad (\text{and hence } i_{o,k} - i_{s,k} < i_o) \\ i_{s,k} \geq i_{o,k} &\Longrightarrow (\beta)_k \quad (\text{and hence } j_{o,k} - j_{s,k} < j_o). \end{aligned}$$

Subcase $a_k + b_k < \min\{i_{o,k} + j_{o,k}, i_{s,k} + j_{s,k}\}$: In this subcase, we have

$$\begin{cases} \nu_{E_{p_k}}(y_1) = a_k + b_k > 0 \\ \nu_{E_{p_k}}(\frac{y_2}{y_1}) = \min\{i_{o,k} + j_{o,k}, i_{s,k} + j_{s,k}\} - (a_k + b_k) > 0, \end{cases}$$

which implies that the generic point of (the strict transform of) E_{p_k} maps under f_1 into the standard neighborhood of q'_1 with a system of regular parameters $(y_1, \frac{y_2}{y_1})$. Therefore, we compute

$$\begin{aligned} \nu_{E_{p_k}}(R_{\log,f_1}) &= \nu_{E_{p_k}}\left(\frac{dy_1}{y_1} \wedge d(\frac{y_2}{y_1}) / \frac{d(\frac{x_{1,k}}{x_{2,k}})}{(\frac{x_{1,k}}{x_{2,k}})} \wedge \frac{dx_{2,k}}{x_{2,k}}\right) \\ &= \nu_{E_{p_k}}\left(\frac{dy_1}{y_1} \wedge dy_2 / \frac{dx_{1,k}}{x_{1,k}} \wedge \frac{dx_{2,k}}{x_{2,k}}\right) - \nu_{E_{p_k}}(y_1) \\ &= i_{o,k} + j_{o,k} - (a_k + b_k). \end{aligned}$$

Now we claim that

$$i_{o,k} + j_{o,k} - (a_k + b_k) < \max\{i_o, j_o\}$$

unless $g_k^{-1}(y_1, y_2) \cdot \widehat{\mathcal{O}_{X_{1,k},p_k}}$ is already principal.

In fact, we have

$$
i_{o,k} + j_{o,k} - (a_k + b_k) = (i_{o,k} - a_k) + (j_{o,k} - b_k)
$$

$$
= \begin{cases} (i_{o,k} - a_k) - (b_k - j_{o,k}) < i_o & \text{if } b_k \geq j_{o,k} \\ (j_{o,k} - b_k) - (a_k - i_{o,k}) < j_o & \text{if } a_k \geq i_{o,k} \\ \text{or} \\ g_k^{-1}(y_1, y_2) \cdot \widehat{\mathcal{O}_{X_{1,k},p_k}} = (x_{1,k}^{a_k} x_{2,k}^{b_k}) \\ \qquad \text{principal if } a_k < i_{o,k} \ \& \ b_k < j_{o,k} \\ \text{(Note that } a_k < i_{s,k} \ \& \ b_k < j_{s,k} \\ \quad \text{by the subcase assumption } a_k + b_k < \min\{i_{o,k} + j_{o,k}, i_{s,k} + j_{s,k}\}.) \end{cases}
$$

Remark that, under the subcase assumption

$$
a_k + b_k < \min\{i_{o,k} + j_{o,k}, i_{s,k} + j_{s,k}\},
$$

when $g_k^{-1}(y_1, y_2) \cdot \widehat{\mathcal{O}_{X_{1,k},p_k}}$ is not principal, we have

$$
\begin{aligned} b_k \geq j_{o,k} &\Longrightarrow (\gamma)_k \quad (\text{and hence } i_{o,k} - a_k < i_o) \\ a_k \geq i_{o,k} &\Longrightarrow (\delta)_k \quad (\text{and hence } j_{o,k} - b_k < j_o). \end{aligned}
$$

Therefore, we conclude

$$
\nu_{E_{p_k}}(R_{\log, f_1}) < \max\{i_o, j_o\}.
$$

Finally, it remains to prove the assertions $(i)_{k+1}$ with $(*)_{k+1}$ and $(ii)_{k+1}$ on $p_{k+1} \in D_{X_{1,k+1}} \subset X_{1,k+1}$, which is a point over p_k and where g_{k+1} is in Subcase $2_{p_{k+1}} 1_q 2$. When we blow up p_k, there appear two points (lying over p_k) where g_{k+1} is in Subcase $2_* 1_q 2$, one having the standard neighborhood with a system of regular coordinates $(\frac{x_{1,k}}{x_{2,k}}, x_{2,k})$ and the other having the standard neighborhood with a system of regular coordinates $(x_{1,k}, \frac{x_{2,k}}{x_{1,k}})$. By symmetry, we have only to check the assertions on p_{k+1} having the standard neighborhood with a system of regular coordinates $(x_{1,k+1}, x_{2,k+1}) = (\frac{x_{1,k}}{x_{2,k}}, x_{2,k})$.

Firstly, by substituting

$$
\begin{cases} x_{1,k} = x_{1,k+1} x_{2,k+1} \\ x_{2,k} = x_{2,k+1} \end{cases}
$$

into $(*)_k$, it is clear that the condition $(i)_{k+1}$ is satisfied with the system of regular coordinates $(x_{1,k+1}, x_{2,k+1})$ providing the coordinate expression $(*)_{k+1}$.

Secondly, we check the condition $(ii)_{k+1}$, assuming that $g_{k+1}^{-1}(y_1, y_2) \cdot \widehat{\mathcal{O}_{X_{1,k+1},p_{k+1}}}$ is not principal.

Observe that

$$
\begin{aligned} a_{k+1} &= a_k, \\ b_{k+1} &= a_k + b_k, \\ i_{o,k+1} &= i_{o,k}, \\ j_{o,k+1} &= i_{o,k} + j_{o,k}, \\ i_{s,k+1} &= i_{s,k}, \\ j_{s,k+1} &= i_{s,k} + j_{s,k}. \end{aligned}
$$

We analyze what happens at the $(k+1)$-the stage in each of the possibilities $(\alpha)_k, (\beta)_k, (\gamma)_k, (\delta)_k$ at the k-th stage.

$(\alpha)_k$: Suppose that we are in the case $(\alpha)_k$.

If $j_{o,k+1} < j_{s,k+1}$, then, since $i_{o,k+1} = i_{o,k} > i_{s,k} = i_{s,k+1}$, we are in the case $(\alpha)_{k+1}$. We also have

$$\begin{aligned} i_{o,k+1} - i_{s,k+1} &= i_{o,k} - i_{s,k} < i_o \\ i_{o,k+1} - a_{k+1} &= i_{o,k} - a_k < i_o. \end{aligned}$$

If $j_{o,k+1} \geq j_{s,k+1}$, then, since $i_{o,k+1} = i_{o,k} > i_{s,k} = i_{s,k+1}$, we have

$$g_{k+1}^* y_2 = v_{k+1} \cdot x_{1,k+1}^{i_{s,k+1}} x_{2,k+1}^{j_{s,k+1}} \quad \text{with} \quad v_{k+1} \in \widehat{\mathcal{O}_{X_{1,k+1},p_{k+1}}}^{\times} \quad \text{unit}.$$

Therefore, the ideal

$$\begin{aligned} g_{k+1}^{-1}(y_1, y_2) \cdot \widehat{\mathcal{O}_{X_{1,k+1},p_{k+1}}} &= (x_{1,k+1}^{a_{k+1}} x_{2,k+1}^{b_{k+1}}, x_{1,k+1}^{i_{s,k+1}} x_{2,k+1}^{j_{s,k+1}}) \\ &= (x_{1,k+1}^{\min\{a_{k+1}, i_{s,k+1}\}} x_{2,k+1}^{\min\{b_{k+1}, j_{s,k+1}\}}) \end{aligned}$$

is principal.

$(\beta)_k$: Suppose that we are in the case $(\beta)_k$.

If $j_{o,k+1} > j_{s,k+1}$, then since $i_{o,k+1} = i_{o,k} < i_{s,k} = i_{s,k+1}$ we are in the case $(\beta)_{k+1}$. We also have

$$\begin{aligned} j_{o,k+1} - j_{s,k+1} &= (i_{o,k} + j_{o,k}) - (i_{s,k} + j_{s,k}) \\ &= (j_{o,k} - j_{s,k}) - (i_{s,k} - i_{o,k}) < (j_{o,k} - j_{s,k}) < j_o \\ j_{o,k+1} - b_{k+1} &= (i_{o,k} + j_{o,k}) - (a_k + b_k) \\ &= (j_{o,k} - b_k) - (a_k - i_{o,k}) < (j_{o,k} - b_k) < j_o. \end{aligned}$$

(Note that under the case $(\beta)_k$ the inequality $a_k \leq i_{o,k}$ would imply that $g_k^{-1}(y_1, y_2) \cdot \widehat{\mathcal{O}_{X_{1,k},p_k}} = (x_{1,k}^{a_k} x_{2,k}^{b_k})$ is principal, being against the assumption. Thus we have $a_k > i_{o,k}$.)

If $j_{o,k+1} \leq j_{s,k+1}$, then since $i_{o,k+1} = i_{o,k} < i_{s,k} = i_{s,k+1}$ we are in the case $(\delta)_{k+1}$, if $g_{k+1}^{-1}(y_1, y_2) \cdot \widehat{\mathcal{O}_{X_{1,k+1},p_{k+1}}}$ is not principal. We also have

$$\begin{aligned} j_{o,k+1} - b_{k+1} &= (i_{o,k} + j_{o,k}) - (a_k + b_k) \\ &= (j_{o,k} - b_k) - (a_k - i_{o,k}) < (j_{o,k} - b_k) < j_o. \end{aligned}$$

(See the note above.)

$(\gamma)_k$: Suppose that we are in the case $(\gamma)_k$.

Since, under the case $(\gamma)_k$,

$$i_{o,k} \leq i_{s,k} \quad \& \quad j_{o,k} \leq j_{s,k},$$

we have

$$\begin{aligned} i_{o,k+1} &= i_{o,k} \leq i_{s,k} = i_{s,k+1} \\ j_{o,k+1} &= i_{o,k} + j_{o,k} \leq i_{s,k} + j_{s,k} = j_{s,k+1}. \end{aligned}$$

Therefore, supposing $g_{k+1}^{-1}(y_1, y_2) \cdot \widehat{\mathcal{O}_{X_{1,k+1},p_{k+1}}}$ is not principal, we are in the case $(\gamma)_{k+1}$. We also have

$$i_{o,k+1} - a_{k+1} = i_{o,k} - a_k < i_o.$$

$(\delta)_k$: Suppose that we are in the case $(\delta)_k$.

Since, under the case $(\delta)_k$,

$$i_{o,k} \leq i_{s,k} \quad \& \quad j_{o,k} \leq j_{s,k},$$

we have

$$\begin{aligned} i_{o,k+1} &= i_{o,k} \leq i_{s,k} = i_{s,k+1} \\ j_{o,k+1} &= i_{o,k} + j_{o,k} \leq i_{s,k} + j_{s,k} = j_{s,k+1}. \end{aligned}$$

Therefore, supposing $g_{k+1}^{-1}(y_1, y_2) \cdot \widehat{\mathcal{O}_{X_{1,k+1},p_{k+1}}}$ is not principal, we are in the case $(\delta)_{k+1}$. We also have

$$\begin{aligned} j_{o,k+1} - b_{k+1} &= (i_{o,k} + j_{o,k}) - (a_k + b_k) \\ &= (j_{o,k} - b_k) - (a_k - i_{o,k}) < (j_{o,k} - b_k) < j_o. \end{aligned}$$

Therefore, Lemma 3.3.6.C.2 is proved.

Now observe that at $p_0 = p$ where $g_0 = f$ is in Subcase $2_{p_0}1_q2$, conditions $(i)_0$ with coordinate expression $(*)_0$ and $(ii)_0$ clearly hold. Therefore, by Lemma 3.3.6.C.2, at any point p_k (over $p_0 = p$) where the morphism g_k is in Subcase $2_{p_k}1_q2$, and where the ideal $g_k^{-1}(y_1, y_2) \cdot \widehat{\mathcal{O}_{X_{1,k},p_k}}$ is not principal, we see that conditions $(i)_k$ with coordinate expression $(*)_k$ and $(ii)_k$ clearly hold, and that

$$\nu_{E_{p_k}}(R_{\log, f_1}) < \max\{i_o, j_o\}.$$

This completes the proof of the second part of statement b).

This completes the local analysis of our algorithm 2.1 in the case where $\dim X = \dim Y = 2$.

§4. Termination of the algorithm in the case where $\dim X = \dim Y = 2$

4.1. **Purpose of this section.** The purpose of this section is, based upon the local analysis in §3, to observe that the logarithmic ramification divisor *decreases* in the process of the algorithm, under some partial order which we introduce among the Weil divisors (on possibly different ambient spaces) and with respect to which the descending chain condition is satisfied. Thus the algorithm must terminate after finitely many steps, ending with the logarithmic ramification divisor being zero, i.e., achieving toroidalization of the given morphism.

4.2. **Partial order among the Weil divisors.** Let $R_1 = \Sigma a_i F_i$ and $R_2 = \Sigma b_j G_j$ be Weil divisors on varieties X_1 and X_2, respectively, where the F_i and G_j are distinct irreducible components with the coefficients in decreasing order, i.e.,

$$\begin{gathered} a_1 \geq a_2 \geq \cdots \geq a_{s-1} \geq a_s \\ b_1 \geq b_2 \geq \cdots \geq b_{t-1} \geq b_t. \end{gathered}$$

Then we define the partial order among the Weil divisors so that

$$R_1 \geq R_2 \ \text{ if } \ (a_1, a_2, ..., a_{s-1}, a_s) \geq (b_1, b_2, ..., b_{t-1}, b_t),$$

where the second inequality is given with respect to the lexicographical order.

It is clear that the set

$$\{(a_1, a_2, ..., a_{s-1}, a_s)\}$$

satisfies the descending chain condition (i.e., there is no infinite strictly decreasing sequence) and that so does the set of Weil divisors (possibly on different ambient spaces).

4.3. **Termination of the algorithm.**

Theorem 4.3.1. *The algorithm 2.1 for toroidalization of a morphism $f : (U_X, X) \to (U_Y, Y)$ in the logarithmic category, where $\dim X = \dim Y = 2$, terminates after finitely many steps.*

More precisely, in the process of the algorithm

$$\begin{array}{ccccccc} (U_X, X) = (U_{X_0}, X_0) & \overset{\text{cp}_1}{\longleftarrow} & (U_{X_0}, X_0) & \overset{\text{cp}_2}{\longleftarrow} & \cdots \overset{\text{cp}_i}{\longleftarrow} & (U_{X_i}, X_i) & \leftarrow \cdots \\ \downarrow{\scriptstyle f = f_0} & & \downarrow{\scriptstyle f_1} & & & \downarrow{\scriptstyle f_i} & \\ (U_X, X) = (U_{X_0}, X_0) & \overset{\text{bp}_1}{\longleftarrow} & (U_{X_0}, X_0) & \overset{\text{bp}_2}{\longleftarrow} & \cdots \overset{\text{bp}_i}{\longleftarrow} & (U_{X_i}, X_i) & \leftarrow \cdots, \end{array}$$

(o) the logarithmic ramification divisor never increases (with respect to the order introduced in 4.2), i.e.,

$$R_{\log, f_i} \geq R_{\log, f_{i+1}},$$

(i) we have no infinite sequence of consecutive blowups

$$\text{bp}_j \circ \text{bp}_{j+1} \circ \cdots \circ \text{bp}_{j+i} \circ \cdots$$

where the centers are only points q of type 2_q,

(ii) every time we have a point q of type 1_q in the center of bp_i, *the logarithmic ramification divisor strctly decreases (with respect to the order introduced in 4.2), i.e.,*

$$R_{\log, f_i} > R_{\log, f_{i+1}}.$$

Therefore, by the descending chain condition on the partial order, the algorithm must terminate after finitely many steps.

Proof.

(o) In each of the subcases from 3.3.1 through 3.3.6, we conclude in the subsection C that

$$\text{either } R_{\log,f_i} = R_{\log,f_{i+1}} \text{ or } R_{\log,f_i} > R_{\log,f_{i+1}}.$$

(i) We may assume $j = 0$ for the proof. Suppose that at the 0-th stage, the center of bp_1 consists only of points q of type 2_q. This means that $f(R_{\log,f_0})$ consists only of points of type 2_q.

Suppose $p \notin R_{\log,f_0}$, i.e., f_0 is toroidal in a neighborhood of p. Then by 3.3.1.B, over a neighborhood of p, the morphism f_1 remains toroidal. That is to say, over this neighborhood of p, there is no intersection with $R_{\log,f_1}$.

Suppose $p \in R_{\log,f_0}$. Then by the local analysis in 3.3.1, we have coordinate expression

$$\begin{cases} f^*y_1 = u \cdot x_1^{a_p} x_2^{b_p} \\ f^*y_2 = v \cdot x_1^{c_p} x_2^{d_p} \end{cases} \text{where } \det \begin{bmatrix} a_p & b_p \\ c_p & d_p \end{bmatrix} = 0 \text{ with units } u, v \in \widehat{\mathcal{O}_{X_0,p}}^{\times},$$

for some systems of regular parameters (x_1, x_2) and (y_1, y_2) of $\widehat{\mathcal{O}_{X_0,p}}$ and $\widehat{\mathcal{O}_{Y_0,p}}$, compatible with the logarithmic structures.

Remark that the matrix

$$\begin{bmatrix} a_p & b_p \\ c_p & d_p \end{bmatrix}$$

is independent of the choice of the systems of regular parameters, up to the column change and row change, and that the set

$$\left\{ \begin{bmatrix} a_p & b_p \\ c_p & d_p \end{bmatrix}; p \in R_{\log,f_0} \right\}$$

is a finite set.

By the local analysis 3.3.1.B, the canonical principalization cp_1 is an isomorphism in a neighborhood of $p \in R_{\log,f_0}$.

We have two possibilities: either

∘ $p \notin R_{\log,f_1}$

or

∘ $p \in R_{\log,f_1}$ where f_1 is in Subcases $* 2_{q_1} *$ at p with the corresponding matrix $\begin{bmatrix} a_p - c_p & b_p - d_p \\ c_p & d_p \end{bmatrix}$ (under the convention $a_p \geq c_p$ and $b_p \geq d_p$), since we assume that $f_1(R_{\log,f_1})$ consists of points q_1 of type 2_{q_1}.

But from this it follows that the length of the sequence could at most be

$$\max\{a_p + b_p + c_p + d_p; p \in R_{\log,f_0}\},$$

proving statement (i).

(ii) Statement (ii) follows from the local analysis in the subsection C of each of the subcases from 3.3.2. through 3.3.6.

Now termination of the algorithm after finitely many steps is an easy consequence of statements (i) and (ii), combined with the descending chain condition on the set of the Weil divisors under the order introduced in 4.2.

4.4. **Corollaries**

Corollary 4.4.1 (Resolution of singularities of morphisms in the logarithmic category in dimension 2). *Let $f : (U_X, X) \to (U_Y, Y)$ be a morphism in the logarithmic category of nonsingular toroidal embeddings with* $\dim X = 2$. *Then there exist sequences of blowups with permissible centers $\pi_X : (U_{X'}, X') \to (U_X, X)$ and $\pi_Y : (U_{Y'}, Y') \to (U_Y, Y)$ such that the induced map $f' : (U_{X'}, X') \to (U'_Y, Y')$ is log smooth.*

Proof.

When $\dim Y = 0$, there is nothing to prove. When $\dim Y = 1$, it follows from 2.2.1 that f is already log smooth. When $\dim Y = 2$, the assertion is a consequence of Theorem 4.3.1.

Corollary 4.4.2 (Toroidalization in dimension 2). *Toroidalization conjecture holds for a dominant morphism $f : X \to Y$ between nonsingular varieties with* $\dim X = 2$.

Proof.

This follows from 1.5 and Corollary 4.4.1.

4.5. **Further remarks.**

Corollary 4.5.1 (Strong factorization of birational maps in dimension 2). *Strong factorization of birational maps in dimension 2 holds: Let $\varphi : X \dashrightarrow Y$ be a proper birational map between nonsingular varieties in dimension 2. Then there exist a sequence of blowups with points as centers*

$$X = X_0 \stackrel{\phi_1}{\leftarrow} X_1 \stackrel{\phi_2}{\leftarrow} \cdots \stackrel{\phi_{l-1}}{\leftarrow} X_{l-1} \stackrel{\phi_l}{\leftarrow} X_l$$

and a sequene of blowdowns

$$X_l = Y_m \stackrel{\psi_m}{\to} Y_{m-1} \stackrel{\psi_{m-1}}{\to} \cdots \stackrel{\psi_2}{\to} Y_1 \stackrel{\psi_1}{\to} Y_0 = Y$$

such that

$$\varphi = \psi_1 \circ \cdots \circ \psi_m \circ \phi_l^{-1} \circ \cdots \circ \phi_1^{-1}.$$

Proof.

The whole point of mentioning this well-known fact is to emphasize the way we prove it, according to the general strategy Step I, which we have now completed proving toroidalization in dimension 2, and step II, which is a consequence of an easy combonatorial fact of geometry of convex bodies in dimension 2 (See 0.4 and Abramovich-Matsuki-Rashid[5]). We remark that the existing proofs for toroidalization use the factorization theorem of proper birational morphisms via Castelnuovo's contractibility of a (-1)-curve.

4.4.2. *With the assumption of properness on f.*

Suppose that in the setting of 2.1.1 for a morphism $f : (U_X, X) \to (U_Y, Y)$ in the logarithmic category, we put the properness assumption on the morphism f.

Then we have much less possibilities in 3.3.5) Subcase 2_p1_q1 and 3.3.6) Subcase 2_p1_q2: Let U_q be an open neighborhood of q with a system of regular parameters (y_1, y_2) with $H_1 = \{y_1 = 0\}$. Then since f is smooth over $U_q - D_Y = U_q - H_1$, Abhyankar's lemma tells us that the normalization $\mu : \widetilde{U_q} \to U_q$ of U_q in the function field $k(X)$ has a system of regular parameters $(\widetilde{y_1}, \widetilde{y_2})$ such that μ can be written in these coordinate systems

$$\begin{cases} \mu^* y_1 = \widetilde{y_1}^e \\ \mu^* y_2 = \widetilde{y_2}. \end{cases} \quad \text{with } \ b > 0.$$

Now the factorization theorem of proper birational morphisms in dimension 2 tells us that $g : f^{-1}(U_q) \to \widetilde{U_q}$ is a sequence of blowups of points.

3.3.5) Subcase 2_p1_q1: Since G_2 is not exceptional for f and hence not for g, we see that there exists a system of regular parameters (x_1, x_2) of $\widehat{\mathcal{O}_{X,p}}$ such that

$$\begin{cases} g^* \widetilde{y_1} = x_1^{a'} x_2 \\ g^* \widetilde{y_2} = x_1 \end{cases}$$

Combining the above two, we conclude

$$\begin{cases} f^* y_1 = x_1^a x_2^b \\ f^* y_2 = x_1 \end{cases} \quad \text{with } \ a = a'e, b = e.$$

3.3.6) Subcase 2_p1_q2: Since both G_1 and G_2 are exceptional for f and hence for g, we see that there exists a system of regular parameters (x_1, x_2) of $\widehat{\mathcal{O}_{X,p}}$ such that we have either

$$\begin{cases} f^* y_1 = x_1^a x_2^b \\ f^* y_2 = x_1^c x_2^d \end{cases} \quad \text{with } \ \det \begin{bmatrix} a & b \\ c & d \end{bmatrix} \neq 0$$

where

$$a \geq c \ \ \& \ \ b \geq d, \text{ or}$$
$$a \leq c \ \ \& \ \ b \leq d.$$

or

$$\begin{cases} f^* y_1 = u \cdot x_1^a x_2^b \\ f^* y_2 = v \cdot x_1^c x_2^d \end{cases} \quad \text{with } \ \det \begin{bmatrix} a & b \\ c & d \end{bmatrix} = 0 \ \text{ and } \ u, v \in \widehat{\mathcal{O}_{X,p}}^{\times} \ \text{ units.}$$

The conclusion is that, under the properness assumption on f, in both 3.3.5) Subcase 2_p1_q1 and 3.3.6) Subcase 2_p1_q2 the ideal $f^{-1}(y_1, y_2) \cdot \widehat{\mathcal{O}_{X,p}}$ is already principal, and hence that the canonical principalization is an isomorphism in a neighborhood of p.

Under the properness assumption on f, no new irreducible components appear in the logarithmic ramification divisor in the process. Moreover, identifying the strict transforms of the irreducible components in the logarithmic ramification divisors, we conclude (via the *usual* order among the Weil divisors)

3.3.1) Subcase $* 2_q * \longrightarrow R_{\log,f_0} = R_{\log,f_1}$
3.3.2) Subcase $1_p 1_q 0 \longrightarrow R_{\log,f_0} = R_{\log,f_1}$
3.3.3) Subcase $1_p 1_q 1 \longrightarrow R_{\log,f_0} > R_{\log,f_1}$
3.3.4) Subcase $2_p 1_q 0$: Impossible
3.3.5) Subcase $2_p 1_q 1 \longrightarrow R_{\log,f_0} > R_{\log,f_1}$
3.3.6) Subcase $2_p 1_q 2 \longrightarrow R_{\log,f_0} > R_{\log,f_1}$.

Therefore, the proof of Theorem 4.3.1 (ii) becomes much simpler under the properness assumption on f.

REFERENCES

[1] S. Abhyankar, *Simultaneous resolutions for algebraic surfaces,* Amer. J. Math **78** (1956), 761-790

[2] S. Abhyankar, *On the valuations centered in a local domain,* Amer. J. Math **78** (1956), 321-348

[3] D. Abramovich and A.J. de Jong, *Smoothness, semistability, and toroidal geometry,* J. Alg. Geom. **6** (1997), 789-801

[4] D. Abramovich and K. Karu, *Weak semistable reduction in characteristic zero,* Invent. Math. **139** (2000), no. 2, 241-273

[5] D. Abramovich,K. Matsuki and S. Rashid, *A note on the factorization theorem of toric birational maps after Morelli and its toroidal extension,* Tohoku Math. J. (2) **51** (1999), no. 4, 489-537

[6] D. Abramovich, K. Karu, K. Matsuki and J. Włodarczyk, *Torification and factorization of birational maps,* J. Amer. Math. Soc. **15** (2002), 531-572

[7] S. Akbulut and H. King, *Topology of algebraic sets,* MSRI publications 25

[8] E. Bierstone and P. Milman, *Canonical desingularization in characteristic zero by blowing up the maximum strata of a local invariant,* Invent. Math. **128** (1997), 207-302

[9] C. Christensen, *Strong domination/weak factorization of three dimensional regular local rings,* Journal of the Indian Math. Soc. **45** (1981), 21-47

[10] S.D. Cutkosky, *Local factorization of birational maps,* Advances in Math. **132** (1997), 167-315

[11] S.D. Cutkosky, *Local monomialization and factorization of morphisms,* Astérisque 260, Soc. Math. France (1999)

[12] S.D. Cutkosky, *Monomialization of morphisms from 3-folds to surfaces,* Springer Lecture Notes in Mathematics, LNM 1786 (2002), Springer-Verlag

[13] S.D. Cutkosky and O. Piltant, *Monomial resolutions of morphisms of algebraic surfaces,* Special issue in honor of Robin Hartshorne, Comm. Algebra **28** (2000), no. 12, 5935-5959

[14] S. Encinas and O. Villamayor, *A course on constructive desingularization and equivariance,* in Resolution of singularities (Obergurgl, 1997), Progress in Math. **181** (2000), Birkhäuser, 147-227

[15] S. Encinas and O. Villamayor, *A new theorem of desingularization over fields of characteristic zero,* preprint (1999)

[16] H. Hironaka, *Resolution of singularities of an algebraic variety over a field of characteristic zero,* Annals of Math. **79** (1964), 109-324

[17] S. Iitaka, *Algebraic Geometry (An introduction to birational geometry of algebraic varieties),* Graduate Texts in Math. 76 (1982), Springer

[18] A.J. de Jong, *Smoothness, semistability, and alterations,* Publ. Math. I.H.E.S. **83** (1996), 51-93

[19] K. Karu, *Local strong factorization of birational maps,* preprint (2003)

[20] K. Kato, *Toric singularities,* Amer. J. Math. **116** (1994), 1073-1099

[21] G. Kempf, F. Knudsen, D. Mumford and B. Saint-Donat, *Toroidal embeddings I,* Lecture Notes in Math. **339** (1973), Springer

[22] H. King, *Resolving singularities of maps,* Real algebraic geometry and topology (East Lansing, Michigan, 1993), Contemp. Math. (1995), Amer. Math. Soc.

[23] K. Matsuki, Correction to "A note on the factorization theorem of birational maps after Morelli and its toroidal extension", Tohoku Math. J. **52** (2000), 629-631

[24] K. Matsuki, *Lectures on factorization of birational maps,* RIMS preprint **1281** (1999)

[25] K. Matsuki, *Notes on the inductive algorithm of resolution of singularities by S. Encinas and O. Villamayor,* preprint (2002)

[26] R. Morelli, *The birational geometry of toric varieties,* J. Alg. Geom. **5** (1996), 751-782

[27] T. Oda, *Convex bodies and algebraic geometry,* Ergebnisse der Mathematik und ihrer Grenzgebiete **15** (1988)

[28] J. Włodarczyk, *Decomposition of birational toric maps in blow-ups and blow-downs (A proof of the weak Oda conjecture),* Transaction of the AMS **349** (1997), 373-411

[29] J. Wlodarczyk, *Birational cobordism and factorization of birational maps,* J. Alg. Geom. **9** (2000), no. 3, 425-449

[30] J. Włodarczyk, *Toroidal varieties and the weak factorization theorem,* preprint math.AG/9904076

[31] O. Zariski, *Algebraic surfaces,* Springer-Verlag (1934)

Contemporary Mathematics
Volume **386**, 2005

Jacobians of quotients of Artin-Schreier curves

Rachel J. Pries

ABSTRACT. We investigate interesting properties of the Jacobian of a universal family of covers $\psi_C : Z_C \to \mathbb{P}^1_C$ over a scheme C in characteristic p. Since we construct this family as a quotient of certain Artin-Schreier curves, the endomorphism algebra of the Jacobian contains many copies of a subfield of $\mathbb{Q}(\zeta_p)$ (which is totally ramified at p). We find the Newton polygon of any geometric fibre of the family using a result of Zhu.

1. Introduction

Arithmetic geometers have been greatly successful at understanding the structure of abelian varieties in characteristic p. For example, the possibilities for the endomorphism algebra and the Newton polygon of a principally polarized abelian variety are known. Furthermore, there are many results on the geometry of the loci of these strata in the moduli space $\mathcal{A}_g$ of principally polarized abelian varieties of dimension g.

These results provide a strong motivation to understand how the Torelli locus intersects such strata in $\mathcal{A}_g$. At this moment, there are many open questions on this subject. In particular, it is unknown which endomorphism algebras and Newton polygons occur for Jacobians of curves in characteristic p. One indication of the difficulty of these questions is to notice the variety of techniques used in the pursuit of examples and partial results. This paper grew as an attempt to understand these questions and techniques in light of a new example.

In Section 2.1, we present a large dimensional family of curves Z_C whose Jacobians have p-rank 0. We show a large ring of endomorphisms acts upon the Jacobian of Z_C in Section 2.2. The main result is Theorem 2.5 which states that $K^r \hookrightarrow \mathrm{End}(\mathrm{Jac}(Z_C)) \otimes \mathbb{Q}$. Here K and r (defined precisely in Section 2.2) are respectively a subfield of $\mathbb{Q}(\zeta_p)$ and a divisor of $p^a - 1$. This result is interesting since by a pure dimension count the Torelli locus and the locus corresponding to abelian varieties with an action of K should not intersect at all in $\mathcal{A}_g$. It is worth noting that the field K is totally ramified above the prime p. This phenomenon occurs since the curves in the family Z_C are quotients of wildly ramified covers of the projective line. In Section 2.3, we compute the $\mathcal{O}_K$-module structure of

1991 *Mathematics Subject Classification.* Primary 14H40; Secondary 14H30, 11G10.
Key words and phrases. Jacobian, endomorphism, Artin-Schreier curve, Newton polygon.

$H^0(\mathrm{Jac}(Z_C), \Omega^1)$. We also use a result in **[Zhu]** to find the Newton polygon of any geometric fibre of Z_C when p satisfies a certain congruence condition.

As an application, in Section 3 we give some results for the Jacobian J of one fibre of the family Z_C. For this fibre, the action of the Cartier operator C is easy to describe. As a result, we find the $\mathcal{O}_K$-module structure of $H^1(J,\Omega^1)/CH^1(J,\Omega^1)$ which determines the a-number and the reduced a-numbers of J. Furthermore, we directly compute the Newton polygon for this fibre of the family. While the result on the Newton polygon follows directly from the fundamental work **[Kat81]** of Katz (or the recent work of Zhu), we hope that this discussion will illuminate some of the issues behind their more technical proofs.

It was a great opportunity to take part in the 2002 Barrett Lectures conference at the University of Tennessee at Knoxville and to hear the plenary lectures by Y. Ihara and Y. Kawamata. This paper fits into the general context of Ihara's work of determining the action of arithmetic Galois groups on covers of the projective line or on the associated Tate modules. To find the action of $\mathrm{Gal}(\mathbb{Q}_p)$, it is valuable to understand the Newton polygon and the endomorphisms of the Jacobians of the covers in question.

I would like to thank Eyal Goren for the opportunity to visit McGill University to study these topics during the summer of 2001.

2. Quotients of Artin-Schreier covers

In this section, we construct a universal family of covers $\psi_C : Z_C \to \mathbb{P}^1_C$ over a scheme C in characteristic p. Then we show that the Jacobians of the curves Z_C in this family have a large endomorphism algebra.

2.1. Construction of the relative curve Z_C. Let k be an algebraically closed field of characteristic $p > 2$. Let $q = p^a$ and choose an isomorphism $(\mathbb{F}_q, +) \simeq (\mathbb{Z}/p)^a$. Fix $n, j \in \mathbb{N}$ such that $n|(p^a - 1)$ and $\gcd(j, pn) = 1$. Consider a semi-direct product $I \simeq (\mathbb{Z}/p)^a \rtimes \mu_n$ where the action of μ_n is irreducible.

Consider a cover $\phi : Y \to \mathbb{P}^1_k$ given by the equations $x_1^n = x$ and $y^{p^a} - y = x_1^j f(x^{-1})$ where $f(x^{-1}) \in k[x^{-1}]$ has degree less than j/n. We will call such a cover an *decomposable I-Galois cover with conductor j*. Here ϕ is an I-Galois cover with Galois action given by $\alpha(y, x_1) = (y + \alpha, x_1)$ for $\alpha \in \mathbb{F}_q$ and $h(y, x_1) = (h^j y, hx_1)$ for $h \in \mu_n$. The cover ϕ has branch locus $\{0, \infty\}$; over 0 it has ramification μ_n and over ∞ it is totally ramified. The filtration of higher ramification groups of ϕ satisfies $I_i \simeq (\mathbb{Z}/p)^a$ for $1 \le i \le j$ and $I_i = 0$ for $i > j$. By the Riemann-Hurwitz formula, the genus g of Y is $g = (p^a - 1)(j - 1)/2$. We assume throughout that $g \ge 2$.

The cover is called *decomposable* since the $(\mathbb{Z}/p)^a$-Galois subcover $\phi' : Y \to X \simeq \mathbb{P}^1_k$ of ϕ can be viewed as the fibre product of related $\mathbb{Z}/p$-Galois covers. More precisely, let L_1 denote a set of representatives of the $(q-1)/(p-1)$ orbits of $\mathbb{F}_q^*$ under the action of $\mathbb{F}_p^*$. For $\ell \in L_1$, let $\phi_\ell : Y_\ell \to X \simeq \mathbb{P}^1_k$ be the $\mathbb{Z}/p$-Galois Artin-Schreier cover $y^p - y = \ell x_1^j f(x^{-1})$. Then ϕ' dominates the cover ϕ_ℓ for $\ell \in L_1$. Thus it dominates the normalization of the fibre product over X of all ϕ_ℓ. Also the genus of Y and of the fibre product are the same. It follows that the normalization of the fibre product of $\{\phi_\ell |\ \ell \in L_1\}$ is isomorphic to ϕ'.

The following lemma indicates that isomorphism classes of decomposable I-Galois covers with conductor j are essentially determined by the choice of $f(x^{-1}) \in$

$k[x^{-1}]$. Furthermore, it states that there is a universal family of such covers whose dimension is $r(I,j)$ which is approximately j/n. Here $r(I,j)$ is the number of exponents $e \in \mathbb{N}$ satisfying: (i) $1 \le e \le j$; (ii) $e \equiv j \bmod n$; (iii) there is no $v \in \mathbb{N}^+$ so that $p^{av}e$ satisfies (i) and (ii). Condition (iii) is important for the following reason: if $g_1(x_1), g_2(x_1) \in k[x_1]$, by [**Pri02**, Lemma 2.1.5] the two covers given by equations $x_1^n = x$ and $y^{p^a} - y = g_i(x_1)$ are isomorphic if and only if $g_1(x_1) = zg_2(x_1) + \delta^{p^a} - \delta$ for some $z \in \mu_{p^a-1}$ and some $\delta \in k[x]$.

Consider the functor $F_{I,j}$ which associates to an irreducible k-scheme S the set of equivalence classes of decomposable I-Galois covers of $\mathbb{P}^1_S$ with conductor j. Here two such covers are *equivalent* if they are isomorphic after a purely inseparable pullback of S. For technical reasons, this functor will only be representable in a category where finite purely inseparable morphisms are invertible. Let $\tau : \phi \to \mathcal{M}_g$ be the forgetful morphism sending ϕ to the isomorphism class of the curve Y.

LEMMA 2.1. *There is a finite quotient $C = C(I,j)$ of $\mathbb{G}_m \times \mathbb{G}_a^{r(I,j)-1}$ by μ_{p^a-1} which is a fine moduli space for the functor $F_{I,j}$ of equivalence classes of decomposable I-Galois covers with conductor j. Thus there is a universal family $\phi_C : Y_C \to \mathbb{P}^1_C$ of equivalence classes of decomposable I-Galois curves with conductor j over $C(I,j)$. The image of ϕ_C under τ in $\mathcal{M}_g$ has dimension $r(I,j)-1$ if $n > 1$ and $r(I,j)-2$ if $n = 1$.*

PROOF. By [**Pri02**, Theorem 2.2.10], in the case $a = 1$, there is a fine moduli space $C = C(I,j)$ for the functor $F_{I,j}$ in a category where finite purely inseparable morphisms are invertible. Here C is constructed as a finite quotient $\mathbb{G}_m \times \mathbb{G}_a^{r(I,j)-1}$ by μ_{p-1}. More specifically: there is a morphism $T : \mathrm{Hom}(\circ, C) \to F_{I,j}(\circ)$ given by using the coordinates of T as the coefficients of the polynomial $f(x^{-1})$; the morphism T induces a bijection between the k-points of C and $F_{I,j}(\mathrm{Spec}(k))$; and if $\phi_S \in F_{I,j}(S)$ then there exists a finite radicial morphism $i : S' \to S$ and a unique morphism $f : S' \to C$ such that $T(f) = i^*\phi_S$ in $F_{I,j}(S')$. The universal family $\phi : Y \to \mathbb{P}^1_C$ is given by $T : \mathrm{Hom}(C,C) \to F_{I,j}(C)$. The proof for $a > 1$ is identical after replacing p with p^a throughout.

By [**Pri02**, Lemma 2.1.2], a subvariety of the family Y_C is isotrivial if and only if it is constant after an automorphism of the projective line which fixes 0 and ∞ if $n > 1$ (resp. fixes ∞ if $n = 1$). Thus the dimension of the image of Y under τ is one (resp. two) less than the dimension of C. □

Note for $n > 1$ that the automorphisms of the projective line which fix 0 and ∞ are only changes in scaling of the variable x, which changes the coefficients of $f(x^{-1})$. So there is a representative of each isomorphism class of the fibres of Y_C for which $f(x^{-1})$ is a monic polynomial. In other words, there is an action of $\mathbb{G}_m$ on C whose orbits correspond to isomorphism classes of the fibres of Y_C.

Suppose $\phi_C : Y_C \to \mathbb{P}^1_C$ is the universal decomposable I-Galois cover with conductor j as above. We will describe ϕ with the equations $x_1^n = x$ and $y^{p^a} - y = x_1^j f(x^{-1})$. Consider the action of a subgroup $H \simeq \mu_n \subset I$ of C-automorphisms on Y_C given by $x_1 \mapsto \zeta_n^i x_1$ and $y \mapsto \zeta_n^{ji} y$ for $0 \le i \le n-1$. The condition that $\gcd(j,n) = 1$ implies that H is not normal in I. Let Z_C be the quotient of the family Y_C by H. Then Z_C is a relative curve over C. Let $\psi_C : Z_C \to \mathbb{P}^1_C$ be the quotient cover which is a non-Galois degree p^a cover of $\mathbb{P}^1_C$ (branched over 0 and ∞).

LEMMA 2.2. *The relative curve Z_C has genus $g' = (q-1)(j-1)/2n$ and p-rank 0. The image of Z_C in $M_{g'}$ has dimension $r(I,j)-1$ (resp. $r(I,j)-2$ if $n=1$).*

PROOF. The subgroup H fixes exactly two points of Y_C, namely the point $\phi_C^{-1}(\infty)$ and the point $(x_1, y) = (0,0)$. One can show that $g' = \text{genus}(Z) = (q-1)(j-1)/2n$ by the Riemann-Hurwitz formula applied to the H-Galois cover $Y_C \to Z_C$.

Recall that $\phi'_C : Y_C \to \mathbb{P}^1_C$ is a p-group Galois cover branched at only one point. Thus the p-rank $\sigma(Y_C)$ of Y_C (and thus Z_C) is 0 by the Deuring-Shafarevich formula: $\sigma(Y_C) - 1 = |(\mathbb{Z}/p)^a|(\sigma(X_C)-1) + \Sigma(e_y - 1)$, [**Cre84**, Corollary 1.8].

The image of Y_C in $\mathcal{M}_g$ has dimension $r(I,j)-1$ (resp. $r(I,j)-2$ if $n=1$). For any irreducible k-scheme S, consider an S-point of C and the corresponding H-Galois cover of relative curves $Y_S \to Z_S$. If Y_S is isotrivial then Z_S is isotrivial since there are only finitely many choices for H in $\text{Aut}_S(Y_S)$, [**DM69**, Theorem 1.11]. If Z_S is isotrivial then Y_S is isotrivial since there are only finitely many μ_n-Galois covers of Z_S branched at two fixed points, [**Gro71**, Exposé X, Corollary 2.12]. So the dimension of the image of Z_C in $M_{g'}$ is the same as the dimension of Y_C in $\mathcal{M}_g$. □

2.2. Endomorphism algebra of the Jacobian. In this section, we show that the endomorphism algebra of the Jacobian of Z_C contains many copies of a subfield of $\mathbb{Q}(\zeta_p)$. In particular, this yields examples of Jacobians which have real multiplication by a totally real field in which the characteristic p ramifies completely. This section is inspired by Ellenberg's paper [**Ell01**] in which he uses I-Galois covers of curves in characteristic 0 to find Jacobians whose endomorphism algebra contains a subfield of $\mathbb{Q}(\zeta_p)$.

Recall the definitions of p, q, n, j, I, H, Y_C, and Z_C from the previous section. Let $n_1 = \gcd(n, p-1)$. Let $K = \mathbb{Q}(\zeta_p^{(n_1)})$ be the subfield of $\mathbb{Q}(\zeta_p)$ which has degree $(p-1)/n_1$ over $\mathbb{Q}$. Note that over the prime p, the field K is totally ramified.

LEMMA 2.3. *Suppose $r = (q-1)/\text{lcm}(n, p-1)$ and $H\backslash I/H$ is the set of double cosets of H in I. There is an isomorphism of $\mathbb{Q}$-vector spaces*

$$\mathbb{Q}[H\backslash I/H] \simeq \mathbb{Q} \oplus r\ \mathbb{Q}(\zeta_p^{(n_1)}).$$

The set of double cosets does not form a group in this case, but Lemma 2.3 shows that $\mathbb{Q}[H\backslash I/H]$ can be given the structure of an algebra.

PROOF. First, we show that the double cosets $H\backslash I/H$ can be identified with the orbits of $\mathbb{F}_q$ under the action of H. Suppose $g_1 = \alpha_1 h_1$, $g_2 = \alpha_2 h_2 \in I$ with $\alpha_i \in \mathbb{F}_q$ and $h_i \in H \simeq \mu_n$. Then $Hg_1H = Hg_2H$ if and only if $\alpha_1 = h(\alpha_2)$ for some $h \in H$. By the Perlis-Walker Theorem [**PW50**], $\mathbb{Q}[\mathbb{F}_q] = \mathbb{Q} \oplus (q-1)/(p-1)\ \mathbb{Q}(\zeta_p)$. If h is a generator of H, then h^{n/n_1} is the stabilizer of $\mathbb{F}_p^* \subset \mathbb{F}_q$. So the double cosets other than H have order $(p-1)/n_1$ in $H\backslash I/H$. Also the number of non-trivial orbits of $\mathbb{F}_q$ under the action of H equals $(q-1)n_1/n(p-1)$, from which the result follows. □

Consider the irreducible representations of I. For each $c \in \mathbb{Z}/n$ there is a representation V_c of I of dimension one which factors through the map $\mathbb{Z}/n \to \mathbb{C}^*$ sending $1 \mapsto e^{2\pi i c/n}$. Let W be the direct sum of the $(q-1)/n$ irreducible representations of I of dimension n. Let W^H be the elements of W fixed by H. So W has dimension $q-1$ and W^H has dimension $(q-1)/n$.

The morphism $i : I \to \mathrm{Aut}(Y_C)$ induces a morphism $i' : \mathbb{Z}[I] \to \mathrm{End}(\mathrm{Jac}(Y_C))$ and defines a morphism $\pi_H = \sum_{h \in H} i'(h) \in \mathrm{End}(\mathrm{Jac}(Y_C))$. Consider the natural action $\mathbb{Q}[H\backslash I/H] \to \mathrm{End}(\pi_H \mathrm{Jac}(Y_C)) \otimes \mathbb{Q}$ and the isogeny $\mathrm{Jac}(Z_C) \to \pi_H \mathrm{Jac}(Y_C)$. Let $\ell \neq p$ be a prime. Then there is an action

$$\mathbb{Q}[H\backslash I/H] \to \mathrm{End}(\mathrm{Jac}(Z_C)) \otimes \mathbb{Q} \hookrightarrow \mathrm{End}(H_1(Z_C, \mathbb{Q}_\ell)).$$

It follows that $H_1(Z_C, \mathbb{Q}_\ell)$ is a $\mathbb{Q}[H\backslash I/H]$-module. Computing the image of $\mathbb{Q}[H\backslash I/H]$ in $\mathrm{End}(\mathrm{Jac}(Z_C)) \otimes \mathbb{Q}$ is equivalent to computing the dimension of the representation of $\mathbb{Q}[H\backslash I/H]$ on $H_1(Z, \mathbb{Q}_\ell) \simeq \pi_H H^1(Y_C, \mathbb{Q}_\ell)$.

PROPOSITION 2.4. *There is a $\mathbb{Q}[H\backslash I/H]$-module isomorphism $H_1(Z_C, \mathbb{Q}_\ell) \simeq (W^H)^{\oplus(j-1)}$.*

In other words, this is an isomorphism of vector spaces which commutes with the action of any $\tau \in \mathbb{Q}[H\backslash I/H]$.

PROOF. Consider the representation ρ_Y determined by the action of I on $H^1(Y, \mathbb{Q}_\ell)$. By the Lefschetz fixed point theorem [**Mil80**, V, Corollary 2.8],

$$\chi(\rho_Y) = 2\chi_{\mathrm{triv}} - 2\chi_1 + a_0 + a_\infty.$$

Here a_0 (respectively a_∞) is the character of the Artin representation attached to the branch point 0 (respectively ∞) and χ_1 (resp. $\chi(\rho_Y)$) is the character of the regular representation (resp. of ρ_Y).

Let $\chi(W)$ be the character associated to the representation W. Then $\chi(W)(g) = 0$ if $g \notin (\mathbb{Z}/p)^a$, $\chi(W)(g) = 0$ if $\mathrm{id} \neq g \in (\mathbb{Z}/p)^a$, and $\chi(W)(\mathrm{id}) = q-1$. Also, $a_\infty(g) = -1$ if $g \notin (\mathbb{Z}/p)^a$, and $a_\infty(g) = -(j-1)$ if $\mathrm{id} \neq g \in (\mathbb{Z}/p)^a$, and $a_\infty(\mathrm{id}) = (p^a - 1)(j-1) + p^a(n-1)$. Also $a_0(g) = 0$ if $g \notin \mathbb{Z}/n$, $a_0(g) = -p^a$ if $\mathrm{id} \neq g \in \mathbb{Z}/n$, and $a_0(\mathrm{id}) = p^a(n-1)$.

Let $m = \langle \chi(W), \chi(\rho_Y) \rangle / \langle \chi(\rho_Y), \chi(\rho_Y) \rangle$. For this m and for some integer m_c for each $c \in \mathbb{Z}/n$, the character $\chi(\rho_Y)$ decomposes as

$$\chi(\rho_Y) = m\chi(W) + \sum_{c \in \mathbb{Z}/n} m_c \chi(V_c).$$

If $c \neq 0$ then only the trivial element of V_c is fixed by H. Thus $H_1(Z, \mathbb{Q}_\ell) \simeq (W^H)^{\oplus m}$.

We compute $m = j-1$ by finding the inner product of $\chi(W)$ with the other characters:

$$\langle \chi(W), \chi(W) \rangle = p(p-1); \quad \langle \chi(W), \chi_{\mathrm{triv}} \rangle = 0; \quad \langle \chi(W), \chi_1 \rangle = np(p-1);$$
$$\langle \chi(W), a_0 \rangle = p(n-1)(p-1); \quad \langle \chi(W), a_\infty \rangle = p(j+n)(p-1).$$

Thus $m = \langle \chi(W), \chi(\rho_Y) \rangle / \langle \chi(W), \chi(W) \rangle = j - 1$. □

It follows that the endomorphism algebra of $\mathrm{Jac}(Z_C)$ contains the direct sum of $(q-1)/\mathrm{lcm}(n, p-1)$ copies of the field K, which has degree $(p-1)/n_1$ over $\mathbb{Q}$ and is totally ramified over the prime p. More precisely:

THEOREM 2.5. *If $r = (q-1)/\mathrm{lcm}(n, p-1)$, then $\oplus_r \mathbb{Q}(\zeta_p^{(n_1)}) \hookrightarrow \mathrm{End}(\mathrm{Jac}(Z_C)) \otimes \mathbb{Q}$.*

PROOF. By Lemma 2.3 and Proposition 2.4, it is sufficient to show that the representation of $(q-1)/(p-1)$ $\mathbb{Q}(\zeta_p)$ on W has dimension $q-1$. This follows since W is the regular representation of $\mathbb{F}_q$ viewed as an I-representation. □

It is natural to ask whether K is the maximal commutative subalgebra in the endomorphism algebra of the generic fibre of $\mathrm{Jac}(Z_C)$. For example, in [**Zar**], Zarhin proves that $\mathbb{Q}(\zeta_n)$ is the maximal commutative subalgebra of the Jacobian for the generic tame cover $y^n = f(x)$ of the projective line.

2.3. Some corollaries. Let L denote a set of representatives of the $(q-1)/\mathrm{lcm}(n, p-1)$ non-trivial orbits of $\mathbb{F}_q$ under the action of H.

COROLLARY 2.6. *There is an isogeny* $\mathrm{Jac}(Z_C) \sim \times_{\ell \in L} J_\ell$. *Here* J_ℓ *is a Jacobian with dimension* $(p-1)(j-1)/2n_1$ *with* $\mathbb{Q}(\zeta_p^{(n_1)}) \hookrightarrow \mathrm{End}(J_\ell) \otimes \mathbb{Q}$. *Furthermore, if* $j = 3$ *and* n_1 *is even (resp. if* $j = 2$ *and* n_1 *is odd) then* J_ℓ *has real (resp. complex) multiplication.*

PROOF. Recall that the normalization of the fibre product of $\{\phi_\ell | \ \ell \in L_1\}$ is isomorphic to ϕ'_C. Here L_1 is a set of representatives of the $(q-1)/(p-1)$ orbits of $\mathbb{F}_q^*$ under the action of $\mathbb{F}_p^*$ and $\phi_\ell : Y_\ell \to \mathbb{P}^1_C$ is the $\mathbb{Z}/p$-Galois cover $y^p - y = \ell x_1^j f(x^{-1})$. The fact that there is an isogeny between $\mathrm{Jac}(Y_C)$ and $\prod_{\ell \in L_1} \mathrm{Jac}(Y_\ell)$ follows from [**KR89**, Theorem B] (or see [**vdG99**, Proposition 2.5]).

Since the action of $H = \langle h \rangle$ on $\mathbb{F}_q$ is irreducible, the orbits of H have order $(q-1)/\mathrm{lcm}(n, p-1)$ and the stabilizer of Y_ℓ is h^{n/n_1}. Define Z_ℓ to be the quotient of Y_ℓ by $\langle h^{n/n_1} \rangle$ and J_ℓ to be the Jacobian of Z_ℓ. After taking the quotient by H, it follows that $\mathrm{Jac}(Z_C) \sim \times_{\ell \in L} J_\ell$; that J_ℓ has dimension $(p-1)(j-1)/2n_1$; and that $\mathbb{Q}(\zeta_p^{(n_1)}) \hookrightarrow \mathrm{End}(J_\ell) \otimes \mathbb{Q}$.

The last statement is immediate since J_ℓ will have real (resp. complex) multiplication if its dimension $(j-1)(p-1)/2n_1$ is the same (resp. half) the degree $(p-1)/n_1$ of $\mathbb{Q}(\zeta_p^{(n_1)})$ (and n_1 is even (resp. odd)). □

REMARK 2.7. a. In the case that $j = 2$, then the family J_ℓ of such curves with complex multiplication has dimension 0. In the case that $j = 3$ and $n = 2$ (resp. $n > 2$) then the family of such curves with complex multiplication has dimension 1 (resp. 0).

b. This result gives some evidence for the remark of Van der Geer and Oort: "one expects excess intersection of the Torelli locus and the loci corresponding to abelian varieties with very large endomorphism rings; that is, one expects that they intersect much more than their dimensions suggest," [**vdGO99**, Section 5]. The Torelli locus has codimension $g(g+1)/2 - (3g-3)$ in $\mathcal{A}_g$. For $a = 1$ and n_1 even, the locus of abelian varieties of dimension $g = (p-1)(j-1)/2n_1$ with endomorphism algebra containing K (which has degree $(p-1)/n_1$) has dimension $(j+1)g/4$. So by a pure dimension count, one would expect these loci not to intersect at all.

c. It is possible to produce other families of curves whose Jacobians have real multiplication. For example, consider a $\mathbb{Z}/p \rtimes \mathbb{Z}/2$-Galois cover $f : V \to \mathbb{P}^1$ whose branch locus consists of two totally ramified points each with conductor j. Then a similar proof shows that $\mathbb{Q}(\zeta_p^{(2)}) \hookrightarrow \mathrm{End}(\mathrm{Jac}(Z)) \otimes \mathbb{Q}$ where Z has genus $j(p-1)/2$. In particular, if $j = 1$ then this family has real multiplication. However, it is less interesting since the p-rank of Z is $(p-1)/2$ and thus Z is ordinary.

COROLLARY 2.8. *Suppose* $p \equiv 1 \bmod j$. *The slopes of the Newton polygon* β *of any geometric fibre of the Jacobian* J_C *of the family* Z_C *are the values* $\{1/j, 2/j, \ldots (j-1)/j\}$ *each with multiplicity* $e = (q-1)/n$. *In other words,* $\beta = e(G_{1,j-1} + G_{2,j-2} + \ldots + G_{j-1,1})$.

PROOF. Since the Newton polygon is invariant under isogeny and $J \sim \times_{\ell \in L} J_\ell$ by Corollary 2.6, it is sufficient to show that the slopes of the Newton polygon of J_ℓ are equally distributed among the values $(j-i)/j$ for $1 \leq i \leq j-1$. By [**Zhu**, Corollary 1.4], the slopes of the Newton polygon of the Jacobian of Y_ℓ are the values $(j-i)/j$ for $1 \leq i \leq j-1$ each with multiplicity $p-1$. The result follows by taking the quotient by the action of H. □

Now we will find the $\mathcal{O}_K$-module structure of $H^0(J_\ell, \Omega^1)$.

LEMMA 2.9. *A basis for* $H^0(J_\ell, \Omega^1)$ *is given by:* $\{y^r x^b dx\}$ *for* $0 \leq r \leq p-2$, $0 \leq b \leq j-2$, $rj + bp \leq pj - j - p - 1$, *and* $jr + b \equiv -1 \bmod n_1$.

PROOF. Let Q be the point of Y_ℓ above $x = \infty$. Note that $(x)_\infty = pQ$, and $(y)_\infty = jQ$, and $(dx) = (2g-2)Q$. So $\operatorname{div}(y^r x^b dx) = r(y)_0 + b(x)_0 + (2g - 2 - (rj + pb))Q$. So the divisors $y^r x^b dx$ for $\operatorname{Jac}(Y_\ell)$ are holomorphic when $rj + pb \leq (2g-2) = pj - j - p - 1$ and are independent for r and b as above.

Recall that $J_\ell = \operatorname{Jac}(Z_\ell)$ where Z_ℓ is the quotient of Y_ℓ by $H_1 \simeq \mu_{n_1}$. Now we find the H_1-invariant subspace of $H^0(\operatorname{Jac}(Y_\ell), \Omega^1)$. The generator ζ_{n_1} of μ_{n_1} acts as follows: $x \mapsto \zeta_{n_1} x$, $y \mapsto \zeta_{n_1}^j y$, $dx \mapsto \zeta_{n_1} dx$. So $y^r x^b dx$ is fixed under the action of μ_{n_1} if and only if $jr + b \equiv -1 \bmod n_1$. □

For $1 \leq i \leq j-1$, let V_i be the subspace of $H^0(J_\ell, \Omega^1)$ generated by the basis elements $\{y^r x^b dx\}$ for which $b = j - 1 - i$. Let $e_i = \#\{r \mid 0 \leq r \leq (pi - j - 1)/j,\ j(r+1) \equiv i \bmod n_1\}$. Note that the dimension of V_i is e_i. When $p \equiv 1 \bmod j$, these numbers simplify to: $r \leq -1 + (j-b-1)(p-1)/j = -1 + i(p-1)/j$ and $e_i = i(p-1)/n_1 j$. Here $n_1 j | (p-1)$ since $\gcd(n, j) = 1$. Let π be a uniformizer for the completion of $\mathcal{O}_K$ above p.

COROLLARY 2.10. *There exists an isomorphism of* $\mathcal{O}_K$*-modules as follows:*

$$H^0(J_\ell, \Omega^1) \simeq \sum_{i=1}^{j-1} k[\pi]/\pi^{e_i}.$$

Note that $H^0(J_\ell, \Omega^1)$ is not a free $\mathcal{O}_K$-module for $j \geq 3$. Thus J_ℓ does not satisfy the Rapoport condition for $j \geq 3$. Also this implies that the canonical principal polarization on J_ℓ is incompatible with the action of $\mathcal{O}_K$.

PROOF. For the proof, we need to determine the action of π on $H^0(J_\ell, \Omega^1)$. We can choose $\zeta_p - 1$ to be a uniformizer for the completion of the ring of integers of $\mathbb{Q}(\zeta_p)$ above p. If N denotes the norm map from $\mathbb{Q}(\zeta_p)$ to K, then $\pi = N(\zeta_p - 1)$ is a uniformizer for the completion of $\mathcal{O}_K$ above p. Note that $\zeta_p - 1$ corresponds to the action of the automorphism $\alpha - \mathrm{id}$ on the function field of Y_ℓ where α is the automorphism of order p acting on Y_ℓ via $\alpha(y) = y + 1$, $\alpha(x) = x$. Thus $\zeta_p - 1$ acts by taking $y^r x^b dx \mapsto ((y+1)^r - y^r) x^b dx$. It follows that $\pi(y^r x^b dx)$ is in the span of the basis vectors $y^{r'} x^b dx$ for which $r' < r$ and $r' \equiv r \bmod n_1$. In particular, the $j-1$ subspaces V_i are the invariant subspaces under the action of π. The value $e_i = \dim(V_i)$ is the power of the uniformizer π which annihilates V_i. □

3. Example

Recall that k is an algebraically closed field of characteristic p, that $n | (p^a - 1)$, and j is a positive integer such that $\gcd(j, pn) = 1$. For the rest of the paper, we suppose that $p \equiv 1 \bmod j$ which implies $\mu_j \subset \mathbb{F}_p^*$. Let Y be the fibre of the decomposable I-Galois family with equations $x_1^n = x$ and $y^{p^a} - y = x_1^j$ (i.e. $f(x^{-1}) = 1$). In this section we restrict to the fibre Z of Z_C, which is the quotient of this fibre Y by the subgroup $H \simeq \mu_n$.

Under these strong conditions, the Jacobian J of Z is easy to investigate since $H^0(J_\ell, \Omega^1)$ decomposes (as an $\mathcal{O}_K$-module) into blocks fixed by the action of the Cartier operator. Here J_ℓ is one of the factors of $J \sim \times_{\ell \in L} J_\ell$ and $\mathcal{O}_K$ is the integral closure of $K = \mathbb{Q}(\zeta_p^{(n_1)})$. We find the $\mathcal{O}_K$-module structure of $H^0(J_\ell, \Omega^1)/CH^0(J_\ell, \Omega^1)$. This allows us to easily compute the Newton polygon of J and the reduced a-numbers of J_ℓ. We show also how the computation of the Newton polygon follows from Katz's fundamental work [**Kat81**].

3.1. Reduced a-numbers. In this section, we describe the action of the Cartier operator C on $H^0(J_\ell, \Omega^1)$ and use this to find the $\mathcal{O}_K$-module structure of $H^0(J_\ell, \Omega^1)/CH^0(J_\ell, \Omega^1)$. Recall that the dimension of J_ℓ equals $(j-1)(p-1)/2$, that the p-rank of J_ℓ is 0, and that J_ℓ is isomorphic to $\mathrm{Jac}(Z_\ell)$ where Z_ℓ is the quotient of Y_ℓ by the action of $H_1 \simeq \mu_{n_1}$ (and $n_1 = \gcd(n, p-1)$).

For b such that $0 \leq b \leq j-2$, let $h_b = (b+1)(p-1)/j$. Note that $n | h_b$ and that h_b is the unique integer in $[0, p-1]$ such that $h_b \equiv (-1-b)j^{-1} \bmod p$.

LEMMA 3.1. *The Cartier operator acts on $H^0(J_\ell, \Omega^1)$ as follows: $C(y^r x^b dx) = 0$ if $r < h_b$ and $C(y^r x^b dx)$ is in the subspace generated by $y^{r-h_b} x^b dx$ if $r \geq h_b$.*

PROOF. Consider the action of the Cartier operator on $y^r x^b dx$.

$$C(y^r x^b dx) = C((y^p - x^j)^r x^b dx) = \sum_{h=0}^{r} C((-1)^h \binom{r}{h} y^{p(r-h)} x^{jh} x^b dx).$$

Thus $C(y^r x^b dx) = \sum_{h=0}^{r} c_{rh} y^{r-h} C(x^{jh+b} dx)$ where $c_{rh} = (-1)^h \binom{r}{h}$ (note $c_{rh} \in \mathbb{F}_p$). Recall that $C(x^{jh+b} dx) = 0$ unless $jh + b \equiv -1 \bmod p$. Since $0 \leq h \leq r \leq p-2$, this is possible if and only if $h = h_b$. Thus $C(x^{jh+b}) \neq 0$ if and only if $h = h_b$. Thus, $C(y^r x^b dx) \neq 0$ if and only if $r \geq h_b$.

When $r \geq h_b$ then $C(y^r x^b dx) = c_{rh} y^{r-h_b} x^{\lfloor (h_b j + b)/p \rfloor} dx$. Also $(h_b j + b)/p = ((b+1)(p-1)+b)/p = b+1-1/p$ so $\lfloor (hj+b)/p \rfloor = b$. So $C(y^r x^b dx) = c_{rh} y^{r-h_b} x^b dx$. □

PROPOSITION 3.2. *Write $a_i = \min\{i, j-i\}(p-1)/n_1 j$ for $1 \leq i \leq j-1$. There exists an isomorphism of $\mathcal{O}_K$-modules as follows:*

$$H^0(J_\ell, \Omega^1)/CH^0(J_\ell, \Omega^1) \simeq \sum_{i=1}^{j-1} k[\pi]/\pi^{a_i}.$$

The numbers a_i are the *reduced a-numbers* of J_ℓ.

PROOF. Recall that $C(y^r x^b dx) = 0$ if $r < h_b$ and $C(y^r x^b dx)$ is in the subspace generated by $y^{r-h_b} x^b dx$ if $r \geq h_j$. Thus the subspaces V_i are invariant under the action of the Cartier operator. Also $\dim(CV_i) = \max\{\dim(V_i) - (j-i)(p-1)/jn_1, 0\}$. As in the proof of Corollary 2.10, the $j-1$ subspaces V_i are the invariant subspaces under the action of π. The value a_i is the power of the uniformizer π which anniliates V_i/CV_i. Since $\dim(V_i) = i(p-1)/n_1 j$, we see that $a_i = \dim(V_i/CV_i) = \min\{i, j-i\}(p-1)/n_1 j$. □

Recall that the *a-number* of an abelian variety A over an algebraically closed field is $\dim_k \mathrm{Hom}(\alpha_p, A)$ where α_p is the group scheme which is the kernel of Frobenius $\mathrm{Fr} : \mathbb{G}_a \to \mathbb{G}_a$ (written $\alpha_p = \mathrm{Spec}(k[x]/x^p)$). By duality, the a-number equals the dimension of the kernel of the Cartier operator on $H^0(J_\ell, \Omega^1)$. The next corollary shows that the a-number of J is approximately half of its genus.

COROLLARY 3.3. *Suppose $p \equiv 1 \bmod j$. The quotient curve Z_ℓ has a-number $(p-1)j/4n_1$ if j is even and $(p-1)(j-1)(j+1)/4jn_1$ if j is odd.*

PROOF. This is immediate from Proposition 3.2 since the a-number is the sum of the reduced a-numbers or $\sum_{i=1}^{j-1} \min\{i, j-i\}(p-1)/jn_1$; this is because it is the dimension of the kernel of the Cartier operator on $H^0(J_\ell, \Omega^1)$. □

REMARK 3.4. Suppose $p \not\equiv 1 \bmod j$. Define h_b to be the unique integer in $[0, p-1]$ such that $h_b \equiv (-1-b)j^{-1} \bmod p$. Then $C(y^r x^b dx) = 0$ if $r < h_b$ and $C(y^r x^b dx) = c_{rh} y^{r-h_{jb}} x^{\lfloor (hj+b)/p \rfloor} dx$ if $r \geq h_{jb}$. So when $p \not\equiv 1 \bmod j$, the proof of Lemma 3.1 gives the following expression for the a-number α_ℓ of the Jacobian J_ℓ in terms of h_b:

$$\alpha_\ell = \sum_{b=0}^{j-2} \min(h_b, p - \lceil (p+1+bp)/j \rceil).$$

This follows since the dimension of the kernel of the Cartier operator on $H^0(J_\ell, \Omega^1)$ is:

$$\#\{(r,b) | 0 \leq r < h_b,\ 0 \leq b,\ rj + bp \leq jp - j - p - 1\}.$$

For example, if $p \equiv -1 \bmod j$ then $h_b = p-(b+1)(p-1)/j$. Thus $C(y^r x^b dx) = 0$ for all $y^r x^b dx \in H^0(J_\ell, \Omega^1)$. Thus $\alpha_\ell = (j-1)(p-1)/2$ and Z is supersingular. When $n_1 = 1$, this is one direction of the main result of [**IS91**]. We note in passing that for $j = q+1$, the curve Y is isomorphic over $\mathbb{F}_{q^2}$ to the maximal curve studied by [**FGT97**]. In other words, the number of $\mathbb{F}_{q^2}$-points of Y is equal to the Hasse-Weil bound.

3.2. Newton polygons. By Corollary 2.8, the slopes of the Newton polygon β of the Jacobian J_C of Z_C are the values in the set $\{1/j, 2/j, \ldots (j-1)/j\}$ each with multiplicity $e = (q-1)/n$. In other words, $\beta = e(G_{1,j-1} + G_{2,j-2} + \ldots G_{j-1,1})$. In this section, we give two explicit proofs of this result for the Jacobian J of the fibre Z of the family Z_C. The first proof considers the action of the Cartier operator on $H^0(J_\ell, \Omega^1)$ and uses Katz's sharp slope estimate. These are also ingredients of Zhu's proof, which is more complicated since the subspaces V_i are not invariant under the Cartier operator and since it is necessary to lift to characteristic 0. The second proof uses a result of Katz, in which he relates the eigenvalues of Frobenius on this fibre to certain Gauss sums. After these proofs we find some lower bounds for the dimension of the intersection of the Torelli locus with certain Newton polygon strata in $\mathcal{A}_g$ (the moduli space of principally polarized abelian varieties of dimension g).

PROOF. (First proof of Corollary 2.8 for the Jacobian J of the fibre Z): Since the Newton polygon is invariant under isogeny and $J \sim \times_{\ell \in L} J_\ell$ by Corollary 2.6, it is sufficient to show that the slopes of the Newton polygon of J_ℓ are equally distributed among the values $(j-i)/j$ for $1 \leq i \leq j-1$. Recall that $\mathrm{End}(J_\ell) \otimes \mathbb{Q}$ contains the field $K = \mathbb{Q}(\zeta_p^{(n_1)})$ and that $\pi = N(\zeta_p - 1)$ is a uniformizer for the completion of K over the prime p.

The subspace V_i of $H^0(Z, \Omega^1)$ is invariant under the action of the Cartier operator. Let $h_i = (j-i)(p-1)/j$. For all $(j-1)/2 \leq i \leq j-1$, we see that h_i/n is the largest exponent ϵ such that $C(V_i) \subset \pi^\epsilon V_i$. This follows since the Cartier operator acts on $y^r x^{j-1-i} dx$ by reducing the exponent r by h_i, by Lemma 3.1. The uniformizer π of $\mathcal{O}_K$ acts by reducing the exponent r by n. Also the Cartier operator has coefficients in $\mathbb{F}_p^*$ since $c_{rh} \in \mathbb{F}_p^*$.

Thus the eigenvalues for C on V_i are divisible exactly by $\pi^{h_i/n}$ for all $(j-1)/2 \leq i \leq j-1$. We see that $\pi^{h_i/n}$ has valuation $v_i = h_i/ne_1 = (j-i)/j$. Thus p^{v_i} exactly divides the Verschiebung endomorphism V on V_i for $(j-1)/2 \leq i \leq j-1$. We can show this implies that $p^{\lceil nv_i \rceil} | V^{n+g}$. By Katz's sharp slope estimate [**Kat79**, Theorem 1.5], the first slope of the Newton polygon on V_i is exactly $(j-i)/j$.

The proof follows from the following two statements. For each slope λ which occurs in the Newton polygon, the slope $1-\lambda$ must occur by duality. Also, (as in [**Yu**, Lemma 3.2]), the slopes must occur with multiplicity at least $e_1 = (p-1)/n_1$ since p is totally ramified in $\mathcal{O}_K$. □

Now here is the proof using Katz's work on eigenvalues of Frobenius and Gauss sums. Note that there is an action of μ_j on Y_ℓ whose generator takes $(x_1, y_\ell) \mapsto (\zeta_j x_1, y_\ell)$.

PROOF. (Second proof of Corollary 2.8 for the Jacobian J of the fibre Z): It is sufficient to show that the slopes of the Newton polygon of the curve $Y_\ell : y^p - y = x_1^j$ are equally distributed (with multiplicity $p-1$) among the values $(j-i)/j$ for $1 \leq i \leq j-1$. The general case follows by Corollary 2.6 and by considering the quotient by the action of the subgroup H.

By [**Kat81**, Corollary 2.2], $H^1(Y_\ell) = \oplus (H^1)^{(\psi,\chi)}$. Here the sum is over all pairs (ψ, χ) where ψ is a nontrivial additive character of $\mathbb{F}_p$ and χ is a nontrivial multiplicative character of μ_j. The statement holds for l-adic or crystalline cohomology groups. Furthermore, for each of the $(j-1)(p-1)$ pairs (ψ,χ), the eigenspace $H^1(Y_\ell)^{(\psi,\chi)}$ has dimension 1. Finally, on $H^1(Y_\ell)^{(\psi,\chi)}$ the eigenvalue of Frobenius is the Gauss sum $-g(\psi,\chi,P)$.

There are $\varphi(m)$ non-trivial characters χ of μ_j having order $m|j$. The choice of such a character is equivalent to a choice of prime P of $\mathbb{Q}(\zeta_m)$ containing p. We identify the numbers $1 \leq t < m$ with $\gcd(t,m) = 1$ with elements $\sigma_t \in \mathrm{Gal}(\mathbb{Q}(\zeta_m)/\mathbb{Q})$. By the Stickelberger relation, [**IR90**, 14, Theorem 2], $g(\psi,\chi,P)^m = P^{\sum t\sigma_t}$ where the sum is over $1 \leq t < m$ with $\gcd(t,m) = 1$. Thus the valuations v_t of $g(\psi,\chi,P)$ for the $\varphi(m)$ characters with order m are the numbers t/m. The result follows since the number of eigenvalues with slope t/m is the number of non-trivial additive characters ψ of $\mathbb{F}_q$, namely $p-1$. □

REMARK 3.5. a. In fact, a third proof of Corollary 2.8 for this fibre should be possible along the following lines. The Artin-Schreier curve $y^p - y = x^j$ is a quotient of the Fermat curve $w^{j(p-1)} + v^{j(p-1)} = 1$, [**Kat81**, pg. 228]. In [**Yui80**, Theorem 2.8], Yui determined the Newton polygon of the Fermat curve. The result would follow by finding which slopes of the Fermat curve match the appropriate characters from the Artin-Schreier curve.

b. We quickly summarize the case when $j = 3$ and $n_1 = \gcd(p-1, n)$ is even. Then J is isogenous to a direct sum of isomorphic Jacobians which have real multiplication by a totally ramified field, $\mathbb{Q}(\zeta_p^{(n_1)})$, by 2.6. Thus the Newton polygon of J can have at most two slopes. If $p \equiv 2 \bmod 3$ then the a-number of

Z is $(p-1)/n$, by 3.3. So J is supersingular and its Newton polygon has slopes 1/2. If $p \equiv 1 \bmod 3$ then the a-number of Z is $2(p-1)/3n$. By Corollary 2.8, its Newton polygon has two slopes: 1/3 and 2/3.

c. For Newton polygons of hyperelliptic curves, see [**Yui78**]. For supersingular curves in characteristic 2, see [**vdGvdV92**].

Fix $e \in \mathbb{N}^+$ and $j \in \mathbb{N}^+$ with $j|e$; let $g = e(j-1)/2$. Let P_{je} be the infinite set of all primes p for which $p \equiv 1 \bmod j$ and for which the quotient $n = (p-1)/e$ is relatively prime to j. This set of primes has Dirichlet density $\varphi(j)/\varphi(ej)$ since it consists of all primes $p \equiv n_0 e + 1 \bmod je$ where n_0 is a unit modulo j; (for these p, one can always choose $a = 1$).

Let $\beta = \beta_{ej}$ be the Newton polygon with slopes $1/j, 2/j, \ldots (j-1)/j$ each repeated e times. In other words, $\beta = e(G_{1,j-1} + G_{2,j-2} + \ldots G_{j-1,1})$. Let NP_β be the locus in $\mathcal{A}_g$ of all principally polarized abelian varieties of dimension g with Newton polygon β. Note that $NP_\beta = \emptyset$ without the hypotheses on j and g.

COROLLARY 3.6. *The intersection in $\mathcal{A}_g$ of the Torelli locus with NP_β has dimension at least $r(I,j) - 2$ for all primes $p \in P_{je}$.*

Here $r(I,j)$ is approximately $2g/(p-1)$.

PROOF. This follows immediately from Corollary 2.8 and Lemma 2.2. □

In some special cases, we can give an even better lower bound for the dimension of the intersection of the Newton polygon strata with the Torelli locus.

COROLLARY 3.7. *Given a genus g, a prime p, and a Newton polygon stratum β in the following table, the dimension d_β of the intersection of the Newton polygon stratum with the Torelli locus is at least the given number.*

g	β	p	$d_\beta \geq$
4	$4G_{1,1}$	$p \equiv 3, 5 \bmod 8$	3
5	$5G_{1,1}$	$p \equiv 11 \bmod 20$	3
6	$6G_{1,1}$	$p \equiv 13 \bmod 24$	3
6	$2G_{1,2} + 2G_{2,1}$	$p \equiv 7, 13 \bmod 18$	6
6	$G_{1,3} + 2G_{1,1} + G_{3,1}$	$p \equiv 5 \bmod 8$	7
7	$7G_{1,1}$	$p \equiv 11, 15, 23 \bmod 28$	2
8	$8G_{1,1}$	$p \equiv 3, 5, 7, 9, 11, 13 \bmod 16$	1
9	$3G_{1,2} + 3G_{2,1}$	$p \equiv 4, 7, 10, 13, 16, 19, 22, 25 \bmod 27$	6
10	$G_{1,4} + G_{2,3} + G_{3,2} + G_{4,1}$	$p \equiv 6, 11, 16, 21 \bmod 25$	9
12	$4G_{1,2} + 4G_{2,1}$	$p \equiv 7, 13 \bmod 18$	3
15	$G_{1,5} + 2G_{1,3} + 3G_{1,1} + 2G_{2,3} + G_{5,1}$	$p \equiv 7, 31 \bmod 36$	8
21	$G_{1,6} + G_{2,5} + G_{3,4} + G_{4,3} + G_{5,2} + G_{6,1}$	$p \equiv 8, 15, 22, 29, 36, 43 \bmod 49$	6

PROOF. Suppose β is one of the Newton polygons above with corresponding genus g. Let j be the reciprocal of the first slope of β. Let $e = 2g/(j-1)$. For the primes above, $p \equiv 1 \bmod j$ and an explicit calculation shows that there exists an a so that $n = (p^a - 1)/e$ is relatively prime to j. By Corollary 2.8, the intersection of β with the Torelli locus is non-empty.

Let ℓ_β be the number of Newton polygons α for which $\beta > \alpha \geq gG_{0,1} + gG_{1,0}$. In fact, ℓ_β is the number of lattice points beneath β with x-coordinate $1 \leq x \leq g$.

Let d_β be the dimension of the intersection of the Newton polygon strata β with the Torelli locus. By purity of the Newton polygon stratification [**dJO00**], $\dim(\mathcal{M}_g) \leq d_\beta + \ell_\beta$. The proof follows by an explicit calculation of ℓ_β in the cases above, which are the only ones for which $\dim(\mathcal{M}_g) - \ell_\beta$ is positive. □

References

[Cre84] R. Crew. Etale p-covers in characteristic p. *Compositio Math.*, 52(1):31–45, 1984.

[dJO00] A. J. de Jong and F. Oort. Purity of the stratification by Newton polygons. *J. Amer. Math. Soc.*, 13(1):209–241, 2000.

[DM69] P. Deligne and D. Mumford. The irreducibility of the space of curves of given genus. *Inst. Hautes Études Sci. Publ. Math. No.*, 36:75–109, 1969.

[Ell01] J. Ellenberg. Endomorphism algebras of Jacobians. *Adv. Math.*, 162(2):243–271, 2001.

[FGT97] R. Fuhrmann, A. Garcia, and F. Torres. On maximal curves. *J. Number Theory*, 67(1):29–51, 1997.

[Gro71] A. Grothendieck. *Revêtements étales et groupe fondamental.* Springer-Verlag, Berlin, 1971. Séminaire de Géométrie Algébrique du Bois Marie 1960–1961 (SGA 1), Dirigé par Alexandre Grothendieck. Augmenté de deux exposés de M. Raynaud, Lecture Notes in Mathematics, Vol. 224.

[IR90] K. Ireland and M. Rosen. *A classical introduction to modern number theory.* Springer-Verlag, New York, second edition, 1990.

[IS91] S. Irokawa and R. Sasaki. A remark on Artin-Schreier curves whose Hasse-Witt maps are the zero maps. *Tsukuba J. Math.*, 15(1):185–192, 1991.

[Kat79] N. Katz. Slope filtration of F-crystals. In *Journées de Géométrie Algébrique de Rennes (Rennes, 1978), Vol. I*, pages 113–163. Soc. Math. France, Paris, 1979.

[Kat81] N. Katz. Crystalline cohomology, Dieudonné modules, and Jacobi sums. In *Automorphic forms, representation theory and arithmetic (Bombay, 1979)*, pages 165–246. Tata Inst. Fundamental Res., Bombay, 1981.

[KR89] E. Kani and M. Rosen. Idempotent relations and factors of Jacobians. *Math. Ann.*, 284(2):307–327, 1989.

[Mil80] J. Milne. *Étale cohomology.* Princeton University Press, Princeton, N.J., 1980.

[Pri02] R. Pries. Families of wildly ramified covers of curves. *Amer. J. Math.*, 124(4):737–768, 2002.

[PW50] S. Perlis and G. Walker. Abelian group algebras of finite order. *Trans. Amer. Math. Soc.*, 68:420–426, 1950.

[vdG99] G. van der Geer. Cycles on the moduli space of abelian varieties. In *Moduli of curves and abelian varieties*, Aspects Math., E33, pages 65–89. Vieweg, Braunschweig, 1999.

[vdGO99] G. van der Geer and F. Oort. Moduli of abelian varieties: a short introduction and survey. In *Moduli of curves and abelian varieties*, Aspects Math., E33, pages 1–21. Vieweg, Braunschweig, 1999.

[vdGvdV92] G. van der Geer and M. van der Vlugt. Reed-Muller codes and supersingular curves. I. *Compositio Math.*, 84(3):333–367, 1992.

[Yu] C. Yu. On reduction of hilbert-blumenthal varieties. preprint.

[Yui78] N. Yui. On the Jacobian varieties of hyperelliptic curves over fields of characteristic $p > 2$. *J. Algebra*, 52(2):378–410, 1978.

[Yui80] N. Yui. On the Jacobian variety of the Fermat curve. *J. Algebra*, 65(1):1–35, 1980.

[Zar] Y. Zarhin. Cyclic covers of the projective line, their Jacobians and endomorphisms. *J. Reine Angew. Math.*

[Zhu] H.J. Zhu. l-functions of exponential sums over one dimensional affinoids, part 1. preprint.

Department of Mathematics, Columbia University, New York, New York 10027

E-mail address: `pries@math.columbia.edu`

Contemporary Mathematics
Volume **386**, 2005

Gluing Schemes and a Scheme Without Closed Points

Karl Schwede

Communicated by Sándor Kovács

ABSTRACT. We first construct and give basic properties of a fibered coproduct in the category of ringed spaces (which is just a particular type of colimit). We then look at some special cases where this actually gives a fibered coproduct in the category of schemes. Intuitively this is gluing a collection of schemes along some collection of other schemes (possibly subschemes). We then use this to construct a scheme without closed points.

1. Introduction

This paper is in essence a look at a naive attempt to glue schemes together. I define first a fibered coproduct in the category of ringed spaces (this is just a special type of colimit). This particular coproduct is the natural generalization of gluing schemes along open subschemes. In the case of gluing only two schemes together, this coproduct is simply the pushout of the topological spaces combined with the appropriate pullback of the rings. This method, as we will see, does not always produce a scheme even when the ringed spaces involved are schemes and the morphisms between them are morphisms of schemes. However we will examine several cases where this colimit is a scheme.

If we glue two affine schemes together using this method and the object along which we are gluing is a closed subscheme of one of the two schemes, the resulting coproduct is in fact an affine scheme (Theorem 3.4). Furthermore since every scheme is a ringed space we see that a fibered coproduct exists in the category of schemes at least in this case. This theorem has many immediate and perhaps unexpected consequences. First it allows us to glue two abstract schemes together along a common closed subscheme without first embedding the schemes in any ambient space. It allows us to contract any closed subscheme of an affine scheme over a field k, to a point. We do this by gluing the closed subscheme we wish to contract to a k-point. In particular, although it is well known that a line on $\mathbb{P}^2_k$

2000 *Mathematics Subject Classification.* 14A15.
Key words and phrases. Algebraic Geometry.

cannot be contracted, we will see explicitly that a line on $\mathbb{A}^2$ can be (even though the resulting scheme will not be noetherian).

A more specialized application of this method is gluing together a finite collection of closed (but not necessarily reduced) points of an affine scheme over a field k. This can be generalized to gluing together a finite collection of closed points in a quasi-projective variety because in that case every finite collection of points is contained in a single affine open subset. Specifically, if we glue two distinct single points of $\mathbb{A}^1_k$ to a single k-point using this method we get a nodal cubic. If we glue a double point $k[x]/(x^2)$ to a single k-point, we get a cuspidal cubic. This method gives us control over how the points are identified and lets us glue abstractly without first mapping the scheme into some projective space.

Another case where 3.4 can be directly applied is if we remove a part of an affine scheme via some localization (inverting certain elements). Then we can glue back a closed subscheme of the original affine scheme to recover some of the points that were removed (those that sat on the closed subscheme we glued back on). These points however have some topological oddities associated with them. In the resulting scheme we can only get to those recovered points along the scheme we glued back. See Proposition 2.6, Corollary 3.13, and Example 3.14. Readers interested in these gluing methods may also want to look at connections to [**1**] (6.1).

Finally as a corollary of 3.4, we give an example of a scheme without closed points. We then look at an alternate construction of the same scheme using valuation rings. This second construction was also independently suggested to Arthur Ogus by Offer Gabber and the details were verified by Bjorn Poonen [**4**], but to my knowledge it has not been published. We also mention results by Mel Hochster [**5**] which imply the existence of a scheme without closed points. The schemes without closed points discussed in this paper are, however, very explicit and constructive.

This paper is self contained. For basic properties of schemes see [**3**]. Basic categorical definitions can be found in [**6**]. Algebraic results including a discussion of the limit and colimit of a diagram of rings can be found in [**2**].

Acknowledgements:
Special thanks to Sándor Kovács

2. A fibered coproduct in the category of ringed spaces and basic properties

We assume that all rings are commutative with unity and maps between rings send 1 to 1.

Suppose $\{X_i\}_{i\in I}$ is a collection of ringed spaces and for each (unordered) pair $i, j \in I$ there exists a ringed space $Z_{i,j}$ and morphisms of ringed spaces $\phi_{(i,j),i} : Z_{i,j} \to X_i$ and $\phi_{(i,j),j} : Z_{i,j} \to X_j$.

DEFINITION 2.1. We define the union of the X_i's along the $Z_{i,j}$'s (which we from now on will denote as $\cup_{Z_{s,t}} X_i$ or when there are only two sets to glue, $X \cup_Z Y$) as the set $\coprod X_i / \sim$ where the relation is generated by relations of the form $x_i \sim x_j$ ($x_i \in X_i$, $x_j \in X_j$) if there exists $z \in Z_{i,j}$ such that $\phi_{(i,j),i}(z) = x_i$ and $\phi_{(i,j),j}(z) = x_j$. Thus two points $x_i \in X_i$ and $x_j \in X_j$ are identified if and only if there exists a finite chain $x_{n_t} \in X_{n_t}$ where $t = 1 \dots m$ with $x_i = x_{n_1}$ and $x_j = x_{n_m}$ and for each pair (n_t, n_{t+1}) we have a $z_{n_t,n_{t+1}} \in Z$ such that $\phi_{(n_t,n_{t+1}),n_t}(z_{n_t,n_{t+1}}) = x_{n_t}$ and $\phi_{(n_t,n_{t+1}),n_{t+1}}(z_{n_t,n_{t+1}}) = x_{n_{t+1}}$ as above. Note that it is possible that $i = j$. We give it the strongest topology such that the natural maps α_s from X_s to $\cup_{Z_{i,j}} X_i$ are all continuous. We will now put a presheaf on the union (we will see later that this is already a sheaf). On each open $U \subset \cup_{Z_{i,j}} X_i$ note that $\alpha_i^{-1}(U)$ is an open subset of X_i and that $\phi_{(i,j),i}^{-1}(\alpha_i^{-1}(U)) = \phi_{(i,j),j}^{-1}(\alpha_j^{-1}(U))$. Define $\Gamma(U, \mathcal{O}_{\cup_{Z_{s,t}} X_i})$ to be the subring of the direct product $\prod_{i \in I} \alpha_i^{-1}(U)$ consisting of all tuples $(a_i)_{i\in I}$ such that $\phi_{(i,j),i}^{\sharp}(a_i) = \phi_{(i,j),j}^{\sharp}(a_j)$ (where in this case the $\sharp$ notation refers to the ring map portion of ϕ) for all pairs $i, j \in I$; in other words, the set of all sections of the X_i that agree on the Z_i. Restriction maps will be simply those induced by the restriction maps on the X_i. If I is empty we define $\cup_{Z_{s,t}} X_i$ as the empty scheme. □

In the case where there are only two X_i's, all we have done here is pushout the topological spaces and pullback the sheaf structure. In the general case this is just a special type of colimit in the category of ringed spaces. We will now show that the above construction is already a sheaf.

PROPOSITION 2.2. *$\cup_{Z_{i,j}} X_i$ is a ringed space.*

Proof: Let $\delta_{i,j} : Z_{i,j} \to \cup_{Z_{i,j}} X_i$ be the composed map $\alpha_i \circ \phi_{(i,j),i} = \alpha_j \circ \phi_{(i,j),j}$ (of topological spaces). Choose an ordering on the set indexing the i's (which index the X_i's). Then we simply note that $\mathcal{O}_{\cup_{Z_{ij}} X_i}$ is just the kernel of the map $\prod (\alpha_s)_* \mathcal{O}_{X_s} \to \prod_{i>j} (\delta_{i,j})_* \mathcal{O}_{Z_{i,j}}$ which sends an element $(\ldots, x_i, \ldots, x_j, \ldots)$ to $(\ldots, \phi_{(i,j),i}^{\sharp}(x_i) - \phi_{(i,j),j}^{\sharp}(x_j), \ldots)$. Kernels of maps are always sheaves, so the presheaf $\mathcal{O}_{\cup_{Z_{i,j}} X_i}$ is a sheaf.

This gives us morphisms of ringed spaces from the X_i to $\cup_{Z_{i,j}} X_i$ (via the α_i) where (on the rings) we simply project to each coordinate. In the same way we get morphisms $\gamma_{i,j}$ from $Z_{i,j}$ to $\cup_{Z_{i,j}} X_i$, those being the composition maps $\alpha_i \circ \phi_{(i,j),i} = \alpha_j \circ \phi_{(i,j),j}$.

This procedure does not in general produce a scheme even if the X_i and $Z_{i,j}$ are schemes. Furthermore, even when it does it may not produce a noetherian scheme even if the all the schemes being glued together are noetherian. The advantage of this definition is that it immediately gives us the following universal property.

THEOREM 2.3. *Suppose that W is a ringed space and there exists maps $\psi_i : X_i \to W$ for each $i \in I$ such that each square of the form*

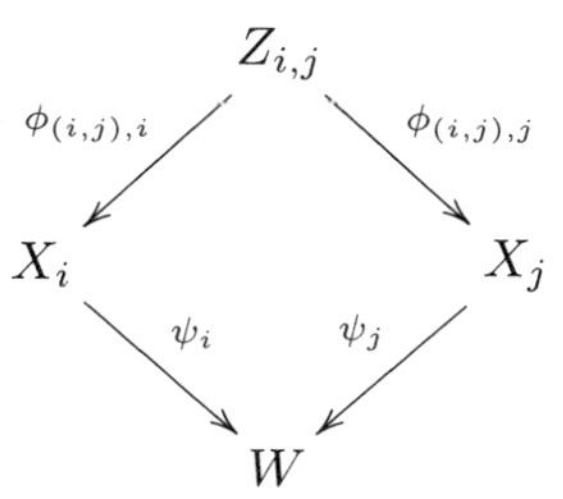

commutes. Then the maps ψ_i factor through the natural maps α_i to $Y = \cup_{Z_{i,j}} X_i$, inducing a unique map $\delta : \cup_{Z_{s,t}} X_s \to W$ so that the following diagrams commute for each $i, j \in I$.

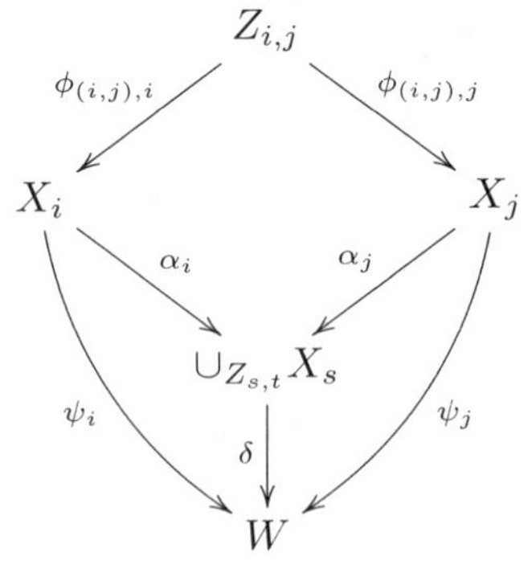

Proof: The proof of this fact is an easy consequence of the definition. □

Let us now record some basic properties of this construction.

LEMMA 2.4. *The open (closed) subsets $\cup_{Z_{i,j}} X_i$, $i \in I$ correspond bijectively to tuples $(U_i)_{i \in I}$ of open (closed) subsets of the X_i such that $\phi^{-1}_{(i,j),i}(U_i) = \phi^{-1}_{(i,j),j}(U_j)$*

Proof: Clearly a subset $U \subset \cup_{Z_{i,j}} X_i$ is open if and only if its pre-image in all the X_i's is open, but it is less clear that given such a tuple (U_i) as described above, there is an open subset that pulls back to it. In fact, we will show that the set $U = \bigcup \alpha_i(U_i)$ has the desired property. So suppose $x_i \in \alpha_i^{-1}(U)$. We need only show that $x_i \in U_i$. Since $\alpha_i(x_i) \in U$, there exists $x_j \in U_j$ with $\alpha_j(x_j)$ identified with x_i in $\coprod X_i$. Without loss of generality let $X_i = X_1$ and $X_j = X_n$ so that there exist $y_t \in X_t$ $(1 \leq t \leq n)$ with $x_i = y_1 \sim y_2 \sim \ldots \sim y_n = x_j$ where for each pair y_t, y_{t+1} we have $z_{t,t+1} \in Z_{t,t+1}$ mapping to y_t and $y_t + 1$ respectively. Note that some of X_t might equal other X_t (ie X_1 might equal X_7). Now by assumption $x_j = y_n \in U_n = U_j$, thus $z_{n-1,n} \in \phi^{-1}_{(n,n-1),n}(U_n)$ so $y_{n-1} = \phi_{(n-1,n),n-1}(z_{n-1,n}) \in \phi_{(n-1,n),n-1}(\phi^{-1}_{(n-1,n),n}(U_n)) \subset U_{n-1}$. Continuing in this fashion we get $x_i = y_1 \in U_1 = U_i$ as desired. □

LEMMA 2.5. *Let $\phi : Z \to X$ and $\psi : Z \to Y$ be morphisms of ringed spaces and let $\alpha : X \to X \cup_Z Y$ and $\beta : Y \to X \cup_Z Y$ be the induced maps. Suppose that ϕ (or ψ) is a homeomorphism onto its image. Then so is β (or α).*

Proof: This follows directly from the universal property.

PROPOSITION 2.6. *Let X, Y and Z be as in the previous lemma with the maps between them labelled in the same way. Now suppose that ψ is a homeomorphism onto its image and that Z is a Zariski space (noetherian and every irreducible closed set has a unique generic point). Then if $x \in X$ and $y \in Y$ and if $\beta(y) \in \{\alpha(x)\}^-$ then there exists $z \in Z$ such that $y \in \{\psi(z)\}^-$ and $\phi(z) \in \{x\}^-$.*

Proof: First note that since ψ is a homeomorphism onto its image, α is also, which implies that $\{x\}^- = \alpha^{-1}(\{\alpha(x)\}^-)$. Now look at $\phi^{-1}(\{x\}^-)$. If this is empty then $\alpha(\{x\}^-) \cup \beta(\emptyset)$ would be closed, in which case $\beta(y) \in \alpha(\{x\}^-)$. However, the points of $\{x\}^- \in X$ cannot be identified with any points of Y since $\phi^{-1}(\{x\}^-) = \emptyset$ which contradicts our assumption $\beta(y) \in \{\alpha(x)\}^-$. So we may assume that $\phi^{-1}(\{x\}^-) \neq \emptyset$. Since ψ is a homeomorphism onto its image, every closed subset of Z arises as the inverse image of a closed subset of Y, (including the inverse images of closed subsets of X). Then because $\beta(y)$ is in the closure of $\alpha(x)$, by

2.4, for every closed subset of $V \subset X$ containing x, each closed subset (there exists at least one) W of Y such that $\psi^{-1}(W) = \psi^{-1}(V)$ contains y. In particular $\psi(\phi^{-1}(\{x\}^-))^-$ contains y. Since Z is Zariski and $\phi^{-1}(\{x\}^-)$ is closed we can write $\phi^{-1}(\{x\}^-) = Z_1 \cup \ldots \cup Z_n$ for irreducible closed sets Z_i with generic points $z_i \in Z_i$. Thus $y \in \psi(Z_1 \cup \ldots \cup Z_n)^- = \psi(Z_1)^- \cup \ldots \cup \psi(Z_n)^-$ which implies that $y \in \psi(Z_i)^-$ for some i. Since continuous maps preserve specialization (points being in the closure of other points) we conclude that $y \in \{\psi(z_i)\}^-$. On the other hand $\phi(z_i) \in \{x\}^-$, which completes the proof. □

We will now prove a result saying we do not have to glue all at once.

PROPOSITION 2.7. *Let X_i and $S = \{Z_{i,j}\}$ be as in definition at the start of the section. Suppose $I = I' \cup I''$ and $I' \cap I'' = \emptyset$. Let S' be the subset of S where both indices of $Z_{i,j}$ are in I' and let S'' be the subset of S where both indices of $Z_{i,j}$ are in I''. Let $S_0 = S - (S' \cup S'')$. Let $Z = \coprod_{S_0} Z_{i,j}$ (as a topological space this is just the disjoint union and on the rings we simply take direct products). Let $X' = \cup_{S'} X_i$ where in this union the X_i are indexed by I'. Likewise let $X'' = \cup_{S''} X_i$. Then there exist maps $\phi' : Z \to X'$ and $\phi'' : Z \to X''$ induced by the $Z_{i,j}$'s making up Z and $X' \cup_Z X'' \cong \cup_{Z_{i,j}} X_i$.*

Proof: This is a direct consequence of the definition or from the fact that colimits commute with colimits. □

Obviously we could take more general partitions as well. This proposition is the computational tool I will use to compute all examples where more than two schemes are glued to together.

Finally I need an algebraic lemma.

LEMMA 2.8. *Finite Limits in the category of commutative rings commute with localization. By this we mean if $\mathscr{B}$ is a finite diagram (a finite number of objects and arrows) in CRng, the category of commutative rings, with limit C, and S is a multiplicative set in C, then $\lim(S^{-1}\mathscr{B}) = S^{-1}C$.*

Proof: Note first that $\mathscr{B}$ can be viewed as a diagram of C algebras and also that $\lim(S^{-1}\mathscr{B})$ is a $S^{-1}C$ algebra by the universal property of the limit [**2**]. Consider now the category of all C algebras. Clearly $C = \lim \mathscr{B}$ in that category as well. Also note that since $\lim S^{-1}\mathscr{B}$ is a $S^{-1}C$ algebra, the limit of that diagram in the category of commutative rings is the same as that in the category of $S^{-1}C$ algebras. We have a functor from C algebras to $S^{-1}C$ algebras which is localization by S, (or tensoring over C by $S^{-1}C$). By Eisenbud [**2**] (A6.1) if we wanted to show that localization commuted with limits, it would be enough show that localization by S commutes with products and equalizers. Since we only need to show it commutes with finite limits, it is enough to show that localization commutes with finite products and equalizers (Eisenbud's proof works in either case).

Clearly localization commutes with finite products (but note that localization does not commute with infinite products). Therefore suppose that that

$$C_1 \underset{\phi_2}{\overset{\phi_1}{\rightrightarrows}} C_2$$

is a diagram of C algebras. Then the equalizer (the limit of that diagram) is just $\{c \in C_1 | \phi_1(c) = \phi_2(c)\}$. This can also be viewed as the kernel K of the map of C

modules $\phi_1 - \phi_2 : C_1 \to C_2$. Likewise the equalizer of the induced diagram

$$S^{-1}C_1 \underset{\overline{\phi_2}}{\overset{\overline{\phi_1}}{\rightrightarrows}} S^{-1}C_2$$

is just the kernel K' of $\overline{\phi_1} - \overline{\phi_2} = \overline{(\phi_1 - \phi_2)} : S^{-1}C_1 \to S^{-1}C_2$. But localization is exact so $S^{-1}K = K'$ as desired.

□

3. An Application to Schemes

We will now use this construction to glue schemes together. Unless otherwise noted, for the rest of the paper X, Y, and Z will be schemes, ϕ will be a map from Z to X and ψ will be a map from Z to Y.

Before we actually do any computations we need to make one more observation. Suppose that we have X, Y and Z as above and there exist open $U \subset X$, $V \subset Y$ such that $W = \phi^{-1}(U) = \psi^{-1}(V)$. Then $U \cup_W V$ is isomorphic to $(X \cup_Z Y)|_{\alpha(U) \cup \beta(V)}$. Topologically this is clear (note what the open subsets of $U \cup_W V$ are). The sections are naturally identified as well.

Example 3.1. (Gluing along open sets) Suppose ϕ and ψ are open immersions. Then $X \cup_Z Y$ is easily seen to be the standard gluing of X and Y along Z and thus a scheme. □

Now let us look at an example where the prime specturms of the pullback in the category of rings is not the corresponding pushout in the category of schemes. We will also see in this example that the pushout in the category of ringed spaces does not produce a scheme.

Example 3.2. Let $X = \operatorname{Spec} k[x, y, y^{-1}]$, $Y = \operatorname{Spec} k[x, y]_{(x,y)}$, and $Z = \operatorname{Spec}(k[x, y]_{(x,y)}[y^{-1}])$. Thus X is the plane without a line, Y is the local ring of the origin, and Z is the local ring of the origin with the line $y = 0$ missing and thus of course the original (closed) point missing as well. Let the maps ϕ and ψ be those induced by inclusions. First note that if one simply pulls back the rings, one gets $k[x, y]$, the spectrum of which is $\mathbb{A}^2_k$. This is not a pushout in the category of schemes because one can certainly factor through $\mathbb{A}^2_k - (1, 0)$ where $(1, 0)$ is the point corresponding to the maximal ideal $(x - 1, y)$. Now let us consider the pushout in the category of ringed spaces. On any open neighborhood of the origin in $X \cup_Z Y$, the corresponding open subset of Y must be all of Y since the only open subset of a local ring that contains the maximal ideal is the entire space. Therefore y must not be invertible in that open set. If we take the spectrum we pick up an open dense subset of the line $y = 0$ in the plane $\mathbb{A}^2$. None of these points (except the origin itself) are in $X \cup_Z Y$, so we do not have a scheme. □

We have just seen how the coproduct in the category of ringed spaces can fail to be a scheme because it has fewer points than the spectrum of sections suggest (even locally). Now let us consider an example where the coproduct in the category of ringed spaces fails to be a scheme because the topology of the coproduct in the category of ringed spaces is too strong to be a scheme, even though all the points are accounted for.

EXAMPLE 3.3. (Gluing along an open/closed subscheme) Suppose $X = \mathbb{A}_k^2 = \operatorname{Spec} k[x,y]$ and $Z = X - \{0\}$ where 0 is the origin. Also let $Y = \mathbb{A}_k^3 - \{0\} = \operatorname{Spec} k[x,y,z] - \{0\}$. Let the maps ϕ and ψ be the natural ones, where ϕ is an open immersion and ψ is a closed immersion. We will show that $X \cup_Z Y$ is not a scheme. Let P be the origin of X (the point missed by ϕ). Each neighborhood of P in $X \cup_Z Y$ corresponds to a pair of open sets $U \subset X$ and $V \subset Y$ such that $\phi^{-1}(U) = \psi^{-1}(V)$. But for any such pair with $P \in U$, V cannot be affine and the prime spectrum of the sections of V will contain the missing origin point. Note that the sections of $U \subset X$ and $\phi^{-1}(U) \subset Z$ are isomorphic for each U so gluing to X neither adds nor removes any new sections besides those associated with Y, i.e. for each open set $U \cup_W V \subset X \cup_Z Y$, the sections $\mathcal{O}_{X \cup_Z Y}(U \cup_W V) = \mathcal{O}_Y(V)$. However, the topology on $X \cup_Z Y$ is too strong for this to be a scheme since not every line (which we think of going through the origin in $\mathbb{A}^3$) actually contains the origin of $X \cup_Z Y$ in its closure as per proposition 2.6. In particular the line Z corresponding to $z = 0$ in $Y \subset \mathbb{A}^3$ does not contain P in it's closure in $X \cup_Z Y$ since $\phi^{-1}(\emptyset) = \psi^{-1}(Z)$. So that the line minus the point is a closed set. □

We have just seen how the topology of the pushout of topological spaces may be stronger than that given by the functions. Now we will give a case where the pushout in the category of ringed spaces is a pushout in the category of schemes. This is the main theorem in the section and all other results in this section follow from it fairly directly. We will also apply it (repeatedly) to produce a scheme without closed points.

THEOREM 3.4. *Suppose A and B are rings. Further suppose I is an ideal of A and there exists a map γ from B to A/I. We will denote the quotient map from A to A/I by π. Let $X = \operatorname{Spec} A$, $Y = \operatorname{Spec} B$ and $Z = \operatorname{Spec} A/I$, so that Z is a closed subscheme of X. Then $X \cup_Z Y$ is an affine scheme with Y a closed subscheme, $(X \cup_Z Y) - Y \cong X - Z$, and the maps $\alpha : X \to X \cup_Z Y$ and $\beta : Y \to X \cup_Z Y$ are morphisms of schemes.*

Informally we are going to glue $Y = \operatorname{Spec} B$ onto X along Z (the map from Z to Y is the one induced by γ). This will replace Z with Y (whatever that might mean) while keeping $X \cup_Z Y$ an affine scheme.

Proof: First let us look at what $X \cup_Z Y$ will look like as a set. Since $\phi : Z \to X$ is a homeomorphism onto its image $\beta : Y \to X \cup_Z Y$ is also. Since Z is a closed subset of X and we are gluing along Z, we see that X must remain the same outside of Z. Furthermore Z will be replaced by Y via the map $\psi : Z \to Y$ induced by γ.

Since I am claiming $X \cup_Z Y$ is affine let us look first at the global sections. The global sections are $C = \{(a,b) \mid a \in A, b \in B, a + I = \gamma(b)\}$. Let the maps $C \to A$, $C \to B$ induced by restricting to each coordinate be denoted by f and g respectively. Let $J = \{(a,0) \in C\} = \ker g$ and let $J' = \{(0,b) \in C\} = \ker f$. It is then easy to see that $C/J \cong B$ with this isomorphism being induced by g. We can view C/J' as a subring of A and in fact it can be thought of as $B[I]$ if A is a B-algebra and π is a map of B-algebras (note A/I is already a B-algebra via γ). In any case we can associate C/J' with $\pi^{-1}(\gamma(B))$. In the future we will denote $\pi^{-1}(\gamma(B)) = \operatorname{im}(C/J') \subset A$ as B'.

Let P be a prime ideal of C. Since $JJ' = (0)$, P must contain either J or J'. If P contains J then P corresponds via g to a prime ideal in B. On the other hand if P does not contain J then it must contain J', so it corresponds to a prime ideal

Q' of B'. Since P didn't contain J, Q' cannot contain I since $J = \{(I,0)\}$. Thus there exists $a \in I$, $a \notin Q'$. Then Q' corresponds to a prime ideal of $B'[a^{-1}] = A$ since $a \in I$. So P corresponds to a prime ideal Q of A. Note that by the naturality of this chain we have $f^{-1}(Q) = P$ and the Q satisfying this property is unique. Thus at least as a set, $\operatorname{Spec} C$ corresponds to $(X - Z) \cup Y$ and the maps f and g induce the expected correspondences.

Now we need to show that this correspondence is in fact a homeomorphism of topological spaces. Let W be a closed subset of the topological space of $X \cup_Z Y$. Thus $\alpha^{-1}(W)$ and $\beta^{-1}(W)$ are closed subsets of X and Y respectively so they are cut out by ideals $K \subset A$ and $K' \subset B$. Let $L = f^{-1}(K) \cap g^{-1}(K') \subset C$. I will show that the points cut out by L are precisely the points of W. First suppose P is a prime ideal of C corresponding to a point of W. Then P comes from a point in X or from a point in Y so it is either a point of $\alpha^{-1}(W)$ or of $\beta^{-1}(W)$. Thus the appropriate corresponding prime of A or B contains either K or K' so it follows that P must contain L. On the other hand, suppose P is a prime ideal of C containing L. Then P must contain $f^{-1}(K)$ or $g^{-1}(K)$. If P contains $g^{-1}(K')$ then it must also contain $J = \ker g \subset g^{-1}(K')$ so that P corresponds to an element of $\beta^{-1}(W)$. If P contains $f^{-1}(K)$ and does not contain J then P corresponds to a prime of A containing K, that is an element of $\alpha^{-1}(W)$. The one case we must worry about is if P contains both $f^{-1}(K)$ and J. In this case I will show that P contains $g^{-1}(K')$ which puts P in the first case again. In particular it is enough to show that $K' \subset g(P)$ since g surjects and P contains $J = \ker g$. Note that $\sqrt{\gamma(K')A/I} = \sqrt{\pi(K)}$ since the inverse images of W in Z must be the same whether we take inverse images through X or Y. Now take $b \in K' \subset B$, then $\gamma(b) \in \gamma(K') \subset \sqrt{\pi(K)}$ so $\gamma(b^n) \in \pi(K)$ for some n. Choose a representative $a + I = \gamma(b^n)$ so that $a \in K$. Then $(a, b^n) \in C$. Since $a \in K$ and $b^n \in K'$ we have $(a, b^n) \subset L$ which implies $(a, b^n) \in P$ so $b^n \in g(P)$ and since $g(P)$ is still prime we conclude $b \in g(P)$ as desired. Therefore, since the topology on $X \cup_Z Y$ was chosen to be the strongest possible, the correspondence we established between $\operatorname{Spec} C$ and $X \cup_Z Y$ is a homeomorphism.

Next we need to show that $X \cup_Z Y$ and $\operatorname{Spec} C$ are isomorphic as schemes. We need only work on the affine open sets $\operatorname{Spec} C_c$ since these sets form a basis. If $c = (s,t) \in C$ then we need to show that the natural map of C_c to the global sections of $(X - V(s)) \cup_{(Z - V(s+I))} (Y - V(t))$ is an isomorphism. But this is just lemma 2.8.

Finally we see that since the maps between the ringed spaces $\alpha : X \to X \cup_Z Y$ and $\beta : Y \to X \cup_Z Y$ are induced by maps of the global sections of these affine schemes we see that α and β are morphisms of schemes as desired. □

THEOREM 3.5. *The scheme constructed in 3.4 is a pushout (fibered coproduct) in the category of schemes.*

Proof: We already know it is a scheme, so we need to show that for any scheme S and for any diagram

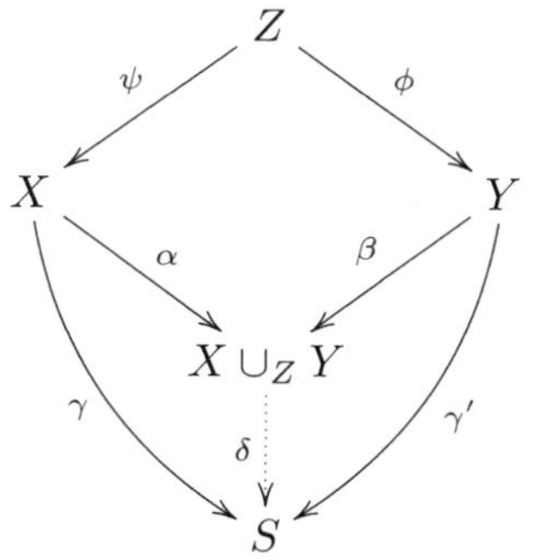

with induced map δ (a map of ringed spaces), that δ is a map of schemes. Note all other maps are maps of schemes. First let $S' \subseteq S$ be an open affine subset. Then $\delta^{-1}(S')$ is open. We can pick an affine open subset of $X \cup_Z Y$ contained in it which is a localization of the scheme $X \cup_Z Y$. We can of course cover $\delta^{-1}(S')$ which such subsets. The corresponding pre-images in X, Y, and Z are localizations also and thus are affine schemes. In this way we may assume without loss of generality that $S = S'$ is affine since a morphism of ringed spaces which is locally a morphism of schemes is a morphism of schemes.

Therefore let $X = \operatorname{Spec} A$, $Y = \operatorname{Spec} B$, $Z = \operatorname{Spec} A/I$, $X \cup_Z Y = \operatorname{Spec} C$, and $S = \operatorname{Spec} R$. We need only show that the map of ringed spaces $X \cup_Z Y \to S$ is induced by $d : R \to C$ (the map on global sections). Let us denote the maps $R \to A$ and $R \to B$ by h and h' respectively. We have the diagram of rings

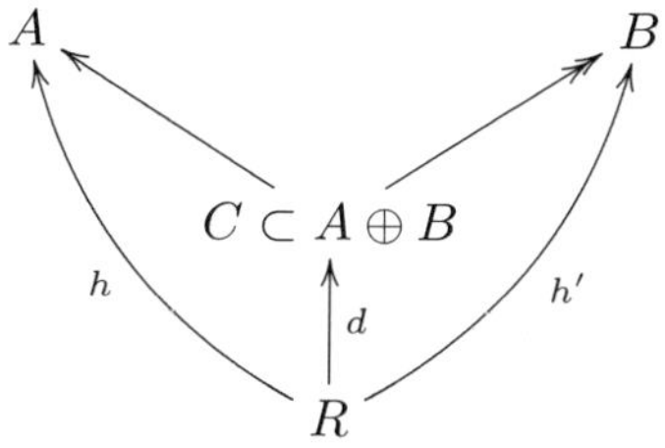

Given a prime $P \subset C$, consider $d^{-1}(P)$. Tracing through the correspondences in the proof of 3.4 we see that $d^{-1}(P)$ is either h^{-1} or h'^{-1} of a prime in A or B respectively such that P is the inverse image. This implies that the map of topological spaces $X \cup_Z Y \to S$ is induced by the map of rings.

To see that the map of ringed spaces really is a map of schemes choose an open affine $\operatorname{Spec} R_r = U \subset S$, where R_r is a localization of R at r. Let $c = (h(r), h'(r)) \in C$. Since we already have the correspondence of topological spaces, $\delta^{-1}(U) = \operatorname{Spec} C_c$. Then we have the above diagram with $A_{h(r)}$, $B_{h'(r)}$, C_c and R_r. All we need to show is that $d : R \to C$ induces $\tilde{d} : R_r \to C_c$ coming from the map of ringed spaces. But $d(x) = (h(x), h'(x))$ so d induces the map $(x/r^n) \mapsto (h(x), h'(x))/c^n = (h(x)/a^n, h'(x)/b^n) = \tilde{d}(x/r^n)$ by the end of the proof of 3.4. Therefore δ is a map of schemes as desired. $\square$

COROLLARY 3.6. *(Contracting a closed set to a point in an affine scheme)* Let A, I, and B be as above and further assume that A is a k algebra and that $B = k$. Then $X \cup_Z Y$ is isomorphic to $\operatorname{Spec} A$ outside of $\operatorname{Spec} A/I$ and the closed subscheme $\operatorname{Spec} A/I$ is contracted to a point.

Proof: Direct from 3.4. □

EXAMPLE 3.7. (Contracting a line in $\mathbb{A}^2_k$). Let $A = k[x, y]$, $I = (x)$, $B = k$, then $X \cup_Z Y = \operatorname{Spec} k[x, xy, xy^2, xy^3, \ldots]$. This is $\mathbb{A}^2_k$ with the line $x = 0$ contracted to a point. Note that in this case the resulting scheme is not noetherian even though the originals were. □

EXAMPLE 3.8. (Gluing points on $\mathbb{A}^1_k$) Let $A = k[x]$ and let $I = ((x-1)(x+1))$; then $X \cup_Z Y$ is easily seen to be isomorphic to the nodal cubic. If $I = (x^2)$ then $X \cup_Z Y$ is the cuspidal cubic. □

We are going to use 3.4 to glue along closed subschemes of arbitrary schemes.

COROLLARY 3.9. *(Gluing closed subschemes in general)* Suppose Z is a closed subscheme of both X and Y. Then $X \cup_Z Y$ is a scheme.

Proof: Outside of Z we must have a scheme, so choose $x \in X$ in the image of Z with $z \in Z$ such that $\phi(z) = x$. Choose an affine open $U \subset X$ with $x \in U$. Then $\phi^{-1}(U)$ is an open affine subset of Z. Since $\psi : Z \to Y$ is a homeomorphism onto its image there exists an open $W \subset Y$ such that $\psi^{-1}(W) = \phi^{-1}(U)$. Choose an affine subset V of W such that $\psi(z) \in V$. Then $\psi^{-1}(V)$ is an affine subset of $\phi^{-1}(U)$. Then there exist further affine subsets (localizations in fact from U and V) $V' \subset V$, $U' \subset U$ such that $z \in \phi^{-1}(U'), \psi^{-1}(V')$ and $\phi^{-1}(U') = \psi^{-1}(V')$. Then $U' \cup_{\phi^{-1}(U')} V'$ is an open neighborhood of x in $X \cup_Z Y$. But by 3.4, that is affine. Thus $X \cup_Z Y$ is locally affine so it is a scheme. □

In fact it is not hard to see that if X and Y are closed subschemes of an ambient scheme and if Z is the scheme corresponding to the intersection (whose ideal sheaf is the sum of X and Y's ideal sheaves) then $X \cup_Z Y$ in fact corresponds to the scheme cut out by the intersection of the ideal sheaf of X with the ideal sheaf of Y. At this point it would be natural to wonder whether we can glue more than two schemes together (perhaps along closed subsets) and still get a scheme. We shall next give an example when this does indeed happen, although not always in the way we expect. First we need a lemma that slightly improves 3.4

LEMMA 3.10. *Suppose $X = \operatorname{Spec} A$ and $Y = \operatorname{Spec} B$ are affine schemes and suppose the scheme $Z = \operatorname{Spec} C$ maps to them via the maps ϕ and ψ as before. Further suppose $\operatorname{im} A \subset \operatorname{im} B \subset C$ and that ψ is a closed map (of topological spaces). Then $X \cup_Z Y$ is an affine scheme with X as a closed subscheme.*

Proof: To prove this we just reduced to the case of 3.4. Let $W = \operatorname{Spec}(\operatorname{im} B)$. Immediately we notice that we have the following diagram.

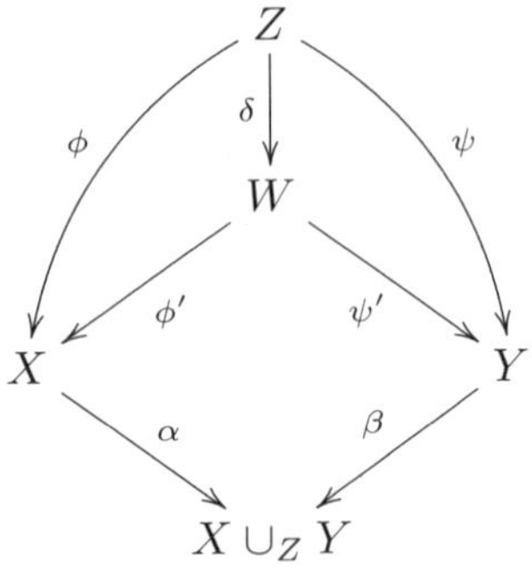

where the map δ is induced by the inclusion of $\operatorname{im} B$ into C, thus W is just the closure of the image of Z in X, but since ψ is closed this is just the image of ψ, and so δ is surjective as well. First let us show that $X \cup_W Y$ and $X \cup_Z Y$ are the same as topological spaces. Since δ is surjective no additional relations are added and by the factorization all original relations are kept; thus $X \cup_W Y$ and $X \cup_Z Y$ are identified as sets. To see that they are identified topologically too, we recall 2.4 and note that if $U \subset X$ and $V \subset Y$ are open subsets such that $\phi^{-1}(U) = \psi^{-1}(V)$ then $\phi'^{-1}(U) = \delta(\phi^{-1}(U)) = \delta(\psi^{-1}(V)) = \psi'^{-1}(V)$ again since δ is surjective. Likewise if $\phi'^{-1}(U) = \psi'^{-1}(V)$ we have that $\phi^{-1}(U) = \delta^{-1}(\phi'^{-1}(U)) = \delta^{-1}(\psi'^{-1}(U)) = \psi^{-1}(U)$. Now we will show that they are isomorphic as sheaves. Choose an open subset of $X \cup_W Y$ corresponding to a pair $U \subset X$, $V \subset Y$. Then the sections of this are the sections that agree in W. But since $\delta^\sharp$ is an injective map of sheaves, these are just the sections that agree in Z as desired. However now we are in the case of 3.4 so $X \cup_Z Y = X \cup_W Y$ is an affine scheme. □

We will now see that if we have a collection of closed subsets of a scheme we can glue them together along their intersections to get a scheme. However, as we will see, this scheme is not always a closed subscheme of the ambient space, although it does always map there (via the universal property).

THEOREM 3.11. *Suppose Y is a scheme and $X_1, \ldots, X_n$ are closed subschemes. Let us denote by $Z_{i,j}$ the closed subscheme corresponding to the intersection of X_i and X_j. Then $\cup_{Z_{i,j}} X_i$ is a scheme and the X_i are closed subschemes.*

Proof: Without loss of generality we may assume that Y is affine (say equal to $\operatorname{Spec} A$) since we can always restrict. Let us denote by I_i the ideal corresponding the X_i so that $Z_{i,j} = \operatorname{Spec}(A/(I_i + I_j))$. We will proceed by induction on n. The base case is clear so suppose we can glue up to n closed subschemes. Let X_{n+1} be another closed subscheme corresponding to an ideal I_{n+1}. Let us denote $X = (\cup_{Z_{i,j},1\leq,i,j\leq n} X_i) = \operatorname{Spec} B$ and $Z = \coprod_{i=1}^n Z_{i,n+1} = \operatorname{Spec} C = \operatorname{Spec} \oplus_{i=1}^n A/(I_i + I_n + 1)$. Then by proposition 2.7 we have $\cup_{Z_{i,j},1\leq,i,j\leq n+1} X_i = X_{n+1} \cup_Z X$. By the universal property 2.3 we see that there are maps $A \to B \to C$ however we also have $A \to A/(I_{n+1}) \to C$ with the first map surjective. Thus $\operatorname{im} A/(I_{n+1} \subset \operatorname{im} B \subset C$ so we can apply lemma 3.10 which shows $\cup_{Z_{i,j}} X_i$ is a scheme. The lemma also guarantees that each X_i is a closed subscheme since the choice of X_{n+1} was arbitrary. □

EXAMPLE 3.12. As noted above, the scheme constructed this way is not always a subscheme of the ambient space (Y in the notation of 3.11). For example let $Y = \mathbb{A}^2 = \operatorname{Spec} k[x, y]$, $X_1 = \operatorname{Spec} k[x, y]/(x)$, $X_2 = \operatorname{Spec} k[x, y]/(y)$, and $X_3 = \operatorname{Spec} k[x, y]/(x - y)$. Now all of the $Z_{i,j}$'s are just $\operatorname{Spec} k = \operatorname{Spec} k[x, y]/(x, y)$. If we glue all three of the lines together simultaneously we get a scheme isomorphic to $\operatorname{Spec} k[x, y, z]/(x, y) \cap (x, z) \cap (y, z)$ which is not equal to $\operatorname{Spec} k[x, y]/(x) \cap (y) \cap (x - y)$ since the dimension of the tangent space at the intersection point of the first is three and the dimension at the intersection point of the second is two. However if we replace the X_i's by schemes with embedded points at the intersection points, that is $X_1 = \operatorname{Spec} k[x, y]/(x^2, xy)$, $X_2 = \operatorname{Spec} k[x, y]/(y^2, xy)$ and $X_3 = \operatorname{Spec} k[x, y]/((x-y)^2, x^2-y^2)$ the scheme obtained by gluing all three simultaneously is $k[x, y, z]/(x) \cap (y) \cap (x - y)$. □

We will now continue on with some other corollaries of 3.4

Corollary 3.13. *(Gluing a closed subset of* Spec R *to* Spec $S^{-1}R$*)* Let R be a ring, S a multiplicative subset of R, and J an ideal of R. Then we let $A = S^{-1}R$, $B = R/J$, and let the ideal I of A be $S^{-1}J$. Then $X \cup_Z Y$ can be identified (as sets) as the union of Spec $S^{-1}R$ and R/J.

Note that while we can make this identification as sets, the topology may be stronger than we might expect. We see this in the next example.

Example 3.14. As in the above corollary let $R = k[x,y]$, $S = \{1, y, y^2, \ldots\}$ and let $I = (x)$. Then $X \cup_Z Y = \operatorname{Spec} k[x, y, \frac{x}{y}, \frac{x}{y^2}, \ldots]$ and this scheme looks like $\mathbb{A}^2$ minus the line $y = 0$ but with the origin put back in (actually the whole line $x = 0$ was put in, but most of it was already there). However, there is a topological pathology created. Proposition 2.6 tells us that the only curves going through the origin must now contain $x = 0$ in their closure. For example the line corresponding to $x - y = 0$ now misses the origin. We can see this algebraically since the ideal $(x - y) = (y)(\frac{x}{y} - 1)$ and $(\frac{x}{y} - 1)$ which is the corresponding prime ideal clearly doesn't go through the origin $m = (x, y, \frac{x}{y}, \frac{x}{y^2}, \ldots) \lhd k[x, y, \frac{x}{y}, \frac{x}{y^2}, \ldots]$. If we took $I = (y)$ instead in an attempt to glue back the line we removed, we notice that Z becomes the empty scheme, so we have only the disjoint union of the plane minus a line and the line. □

Let us conclude with another example which illustrates the gluing procedure of 3.4.

Example 3.15. In this example let $X = \mathbb{A}^2_k = \operatorname{Spec} k[x,y]$, $Y = \mathbb{A}^1_k = \operatorname{Spec} k[x^2]$, and $Z = \operatorname{Spec} k[x,y]/(y) = \operatorname{Spec} k[x] = \mathbb{A}^1_k$. Let all the maps be those induced by the obvious maps of rings ($k[x^2] \subset k[x]$ and $k[x,y] \to k[x,y]/(y)$). Then $X \cup_Z Y$ is a rational surface with the line $y = 0$ folded over on itself. The global sections are just $k[y, xy, x^2] = k[a,b,c]/(a^2c - b^2)$ which is a pinch point singularity. □

4. A scheme without closed points

In this section we will give two constructions of a scheme without closed points. We will utilize the gluing methods from the previous section, the other will use valuation rings. The two schemes we get in this section are easily seen to be the same.

Proposition 4.1. *If X is a quasi-compact scheme then X has a closed point.*

Proof: Since X is quasi-compact there is a cover by affine $\{U_i\}$, $U_i = \operatorname{Spec} A_i$. Take a maximal ideal (closed point) P_1 of U_1. If P_1 is closed in X we are done. If not, take P_2 to be any point (besides P_1 itself) in its closure. Now P_2 is in some U_i (but not U_1) so without loss of generality say P_2 is in U_2. If P_2 is closed we are done. If not take a point P_3 in its closure. Again P_3 is in one of the U_i's but this time it cannot be in either U_1 or U_2 (since it is in the closure of both P_1 and P_2), so without loss of generality we say $P_3 \in U_3$. This process must stop since there are only finitely many U_i, so X has a closed point. □

First we will construct a scheme without closed points using 3.4 and 2.7.

Theorem 4.2. *Let $A_n = (k(x_{n+1}, x_{n+2}, \ldots))[x_n]_{(x_n)}$. Note that this is a DVR with generic point $P_{n-1} = Z_{n-1,n} = \operatorname{Spec} k(x_n, x_{n+1}, \ldots)$ and closed point $P_n =$*

Spec $k(x_{n+1}, x_{n+2}, \ldots)$. Let $X_n = \operatorname{Spec} A_n$. Note that X_n's closed point is identified with X_{n+1}'s generic point so the $Z_{i,j}$ notation is justified. Let all other $Z_{i,j}$'s be the empty scheme. Let us denote the ringed space $\cup_{Z_{i,j}} X_n$ by X. Then X is a scheme.

We will prove this by repeated application of theorem 3.4.

Proof: First notice that what we are doing is gluing DVR's end to end (closed point to open point) in infinite succession. What we have is a succession of points $P_0, P_1, P_2, \ldots$ with each P_i in the closure of all previous P's. Thus clearly X has no closed points. Thus the only open sets (besides the whole set) are finite sets of the form $Y_n = \{P_0, P_1, \ldots, P_n\}$. I claim these finite open sets are affine schemes. We will proceed by induction. Note that the inverse image of Y_n in all but the first n X_i's are empty (therefore the inverse image of the Y_n in higher $Z_{i,i+1}$'s are empty as well), so we can essentially ignore them. Further this means that the sections we get from them are only zero sections. A sort of minimal diagram of this situation is the following.

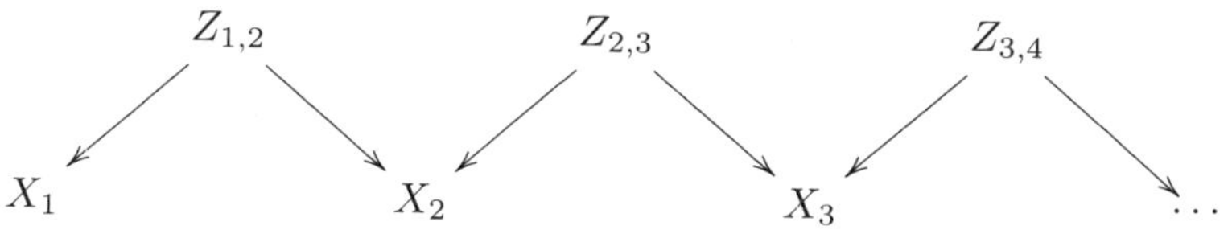

Since these sets are all finite, we can get a perhaps clearer picture of what is going on by simply denoting them by their points.

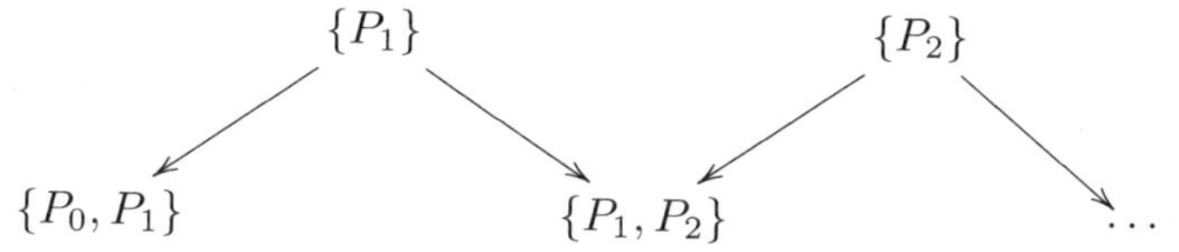

Obviously each pair of P_i's is identified in the coproduct giving the chain of points where each P_{i+1} is in the closure of P_i. Now consider $Y_1 = \{P_0, P_1\}$. This is clearly a scheme because it is the spectrum of $k(x_2, x_3, \ldots)[x_1] = A_1$. So suppose Y_i is a scheme for up to Y_n. Consider Y_{n+1}. By 2.7 we can construct Y_{n+1} by first gluing together Y_n and gluing on the rest. Our induction hypothesis tells us that Y_n is an affine scheme. Let us look at the setup from 2.7. The schemes $X' = Y_n$ and $X'' = X_{n+1}$ are both affine. The Z object from 2.7 is simply $Z_{n-1,n}$ since all the other possible $Z_{i,j}$'s are empty. But the map from Y_n to $Z_{n-1,n}$ is a closed immersion so we can apply 3.4 which tells us that $Y_n \cup_{Z_{n-1,n}} X_{n+1} = Y_{n+1}$ is a scheme. Therefore all Y_n's are schemes and since the collection of all such finite open Y_i's covers X, X is also a scheme. Note that X has no finite affine cover. $\square$

Now we will present an alternate view using valuation rings.

Let

$$A' = k[x_1, x_2, \ldots]\left[\frac{x_1}{x_2}, \frac{x_1}{x_2^2} \cdots\right]\left[\frac{x_2}{x_3}, \frac{x_2}{x_3^2} \cdots\right] \cdots$$

Now the principle of proposition 2.6 suggests that the local ring at the origin will be a local ring with no largest prime ideal among the non-maximal prime ideals. Let $A = A'_{(x_1, x_2, x_3, \ldots)}$ (of course the ideal $\mathfrak{m} = (x_1, x_2 \ldots)$ contains all the monomials of A'). A is the set of global sections of the scheme without closed points constructed above. Note that in A or A' I will call finite products of the generators of A (the $\frac{x_i}{x_{i+1}^n}$) monomials. If $n < m$ I will denote $\frac{x_n}{x_{n+1}}\frac{x_{n+1}}{x_{n+2}}\cdots\frac{x_{m-1}}{x_m^l}$ by $\frac{x_n}{x_m^l}$.

When we take the prime spectrum of A, we will get an infinite chain of points and then a closed point (corresponding to the maximal ideal). If we remove that closed point we get a scheme without closed points.

We are going to see that A is in fact a valuation ring with valuation to $G = \mathbb{Z} \oplus \mathbb{Z} \oplus \ldots$ which gives it properties very much like those of a DVR and actually makes the analysis surprisingly easy. The ordering we are going to use on G is the lexicographic order, $(n_1, n_2, \ldots) > (n_1', n_2', \ldots)$ if the first nonzero entry of $(n_1 - n_1', n_2 - n_2', \ldots)$ is greater than 0. Given $g \in G$ let $lv(g)$ be the value of the leading term of g, let $li(g)$ be the index of the leading term of g, and let g_i denote the i'th entry of g. So for example if $g = (0, 0, 4, -2, \ldots)$ then $lv(g) = 4$ while $li(g) = 3$ (the first two entries are zero). First we will define the valuation v on the monomials of A (or A'). Let m be a monomial of A, and view m as an element of Frac A, then we define $v(m)$ to be the degree (positive or negative) of x_i in m. So for example $v(x_1 \frac{x_1}{x_2^2}\frac{x_2}{x_3}) = (2, -1, -1, 0, 0, 0, \ldots)$. We shall now prove a number of quick results which will allow us to conclude that A is the desired ring.

PROPOSITION 4.3. *For every $g \geq (0, 0, \ldots)$ there exists a unique monic monomial $x \in A$ (or A') such that $v(x) = g$.*

Proof: Let $l = li(g)$ and $n_l = lv(g)$. Note that without loss of generality we can assume $n_l = 1$ for if not we can find an x such that $v(x)$ matches g except at the leading term. That is $lv(v(x)) = 1$ and $v(x x_l^{lv(g)-1}) = g$. By the same method we can assume that $g_i \leq 0$ for all $i > l$ Let $t = \sum_{i>l} \mid g_i \mid$. Then let $x = \frac{x_l}{x_{l+1}^t} \prod_{i>l} \frac{x_{l+1}^{|g_i|}}{x_i^{|g_i|}}$ and note that it satisfies the desired condition. It is unique because monomials of Frac A satisfying that property are unique. □

We will now show that every element of A is a unit times a monomial but first we need a very important lemma.

LEMMA 4.4. *If m_1 and m_2 are monomials of A and if $v(m_1) > v(m_2)$ then m_2 divides m_1.*

Proof: Let m_3 be the monomial corresponding to $v(m_1) - v(m_2)$, then $\lambda m_2 m_3 = m_1$ for some constant $\lambda \in k$ □

PROPOSITION 4.5. *Every element $f \in A$ that is not a unit (so it has no constant term) is a unit multiplied by a unique monic monomial.*

Proof: Represent f as $(\lambda_1 m_1 + \ldots + \lambda_n m_n)/h$ where the $\lambda_i \in K$, $m_1 > m_2 > \ldots > m_n$ and h is an element of $A' - \mathfrak{m}$. Let $m_i' = \frac{m_i}{m_n} \in A' \subset A$. Thus $f = (\lambda_1 m_1' m_n + \ldots \lambda_n m_n)/h = m_n(\lambda_1 m_1' + \ldots \lambda_n)/h$ which is a monic monomial multiplied by a unit as desired. The fact that it is unique is easy to see since it is clear that any two distinct monomials are not associates (they do not differ by a unit). □

We can immediately conclude that every ideal of A is monomial. Now we have the required machinery to actually identify the prime ideals of A and use this to construct a scheme without closed points.

THEOREM 4.6. *With A as described above,* Spec $A - \mathfrak{m}$ *is a scheme without closed points.*